Routledge Revivals

How Animals Develop

First published in 1935 (this edition in 1946), this short account of the science of embryology was the first book in English to provide a simple outline of the whole of this important subject. The study of development is perhaps the best method of approach to the most fundamental of all biological problems, the problem of how all the diverse activities are integrated so as to make up a complete individual organism. The book gives a short sketch of the general pattern on which all animals are built, but devotes more attention to the factors which cause the development of the elements in the pattern, and which then bring them into correct relations with one another.

This volume is simply written in order to enable the general reader to understand the revolutionary advances made in the subject at that time.

I0042194

How Animals Develop

C. H. Waddington

Routledge
Taylor & Francis Group

First published in 1935
This edition first published in 1946
by George Allen and Unwin LTD

This edition first published in 2016 by Routledge
2 Park Square, Milton Park, Abingdon, Oxon, OX14 4RN
and by Routledge
711 Third Avenue, New York, NY 10017

Routledge is an imprint of the Taylor & Francis Group, an informa business

Publisher's Note
The publisher has gone to great lengths to ensure the quality of this
reprint but points out that some imperfections in the original copies may
be apparent.

Disclaimer
The publisher has made every effort to trace copyright holders and
welcomes correspondence from those they have been unable to contact.

A Library of Congress record exists under LC control number: 36018536

ISBN 13: 978-1-138-95668-1 (hbk)
ISBN 13: 978-1-315-66556-6 (ebk)
ISBN 13: 978-1-138-95673-5 (pbk)

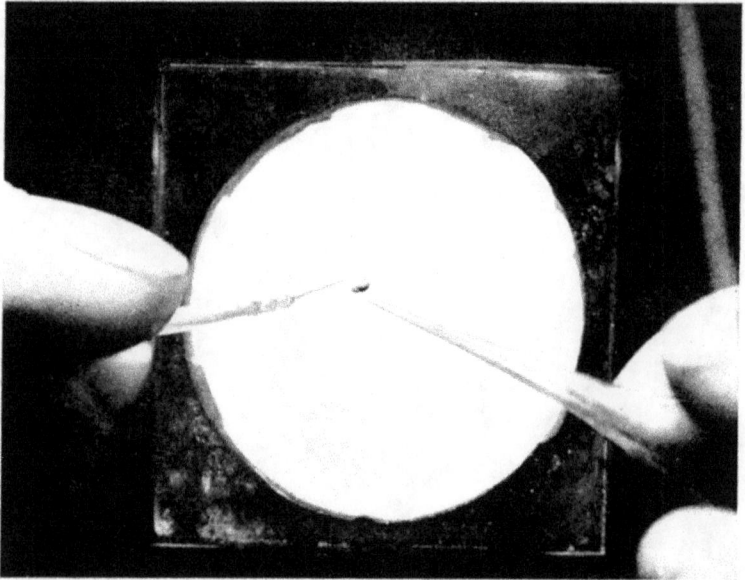

FRONTISPIECE.—Operating on a newt's embryo. The embryo lies in a glass dish lined with wax. The operator has a glass needle in the left hand and in the right hand a hair-loop on a glass-holder; the loop can just be seen crossing the embryo

HOW ANIMALS DEVELOP

by

C. H. WADDINGTON, M.A.

Fellow of Christ's College, Cambridge

*Strangeways Research Laboratory, Laboratory of
Experimental Zoology, Cambridge*

LONDON
GEORGE ALLEN & UNWIN LTD
MUSEUM STREET

FIRST PUBLISHED IN 1935
SECOND IMPRESSION 1940
THIRD IMPRESSION 1946

PRINTED IN GREAT BRITAIN BY
BRADFORD & DICKENS, LONDON, W.C.1

ACKNOWLEDGMENTS

FOR permission to reproduce illustrations acknowledgment is made to the Cambridge University Press (Engelbach, *Endocrinology*; Behrens and Barr, *Endocrinology*; Huxley and de Beer, *Experimental Embryology*), Messrs. Longmans, Green & Co. (Quain, *Anatomy*), Messrs. Macmillan & Co. (MacBride, *Invertebrate Embryology*), the *Railway Gazette* (photograph of Whitemoor Marshalling Yard, L.N.E.R.), and to my own publishers (Dürken, *Experimental Analysis of Development*; Stockard, *Physical Basis of Personality*).

FOREWORD

IN THIS book, I have tried to write an account of
embryology suitable for the intelligent layman and
the elementary student. I have been conscious of
two main difficulties in this task. Firstly, embryos
are complicated and unfamiliar things, so that one
has to describe their structure before one can
discuss the problems they present. I have attempted
to avoid too much description by concentrating on
the early stages of development, which are as a
matter of fact the most important from a general
theoretical point of view, and can be described
fairly shortly, since the embryos have not yet had
time to develop any great complexity of form. The
second difficulty arises because embryology is so
interesting. The development of the structures by
which living things carry out the activities of life
must clearly raise many of the most fundamental
problems about the nature of life itself. But most of
the answers to these problems are still obscure. In
order to show the directions in which people's
thoughts are being led by the recent progress of
embryology, I have put forward some of my own
views, perhaps without sufficient warning that they
represent probabilities rather than certainties. If I
had attempted to give all the possible interpretations
of the facts the book would have become unwieldy
and confusing, but it was impossible to shirk a
discussion of the problems. I believe the ideas which

I have put forward are those most generally held by people who are working at embryology at the present day, but to-morrow we may discover some new fact which will force us to modify them. When one is brought face to face with the most fundamental questions about living things, one cannot expect to obtain complete answers in the comparatively short time during which biology has been actively studied.

C. H. W.

CAMBRIDGE

1935

CONTENTS

LIST OF ILLUSTRATIONS

HOW ANIMALS DEVELOP

CHAPTER I

INTRODUCTION

The Development of Animal Organization

LIVING animals are constantly on the move. It is
one of the most characteristic things about them.
Often we can see them running about, breathing,
catching food and eating it, and so on. If we look
closer we find that an animal is made up of different
organs, and in all of them there is something going
on all the time. On an even smaller scale, the organs
are built out of cells, little lumps of living matter,
each containing a special kernel or *nucleus*. And each
cell is always full of activity. In plants the living
jelly streams slowly about from one side of the cell
to the other: in animal cells we cannot usually see
any movement, but nevertheless there are incessant
chemical actions and reactions. The cell absorbs
oxygen and other substances from outside, performs
many complicated chemical operations with them,
and pours out again into its surroundings the by-
products for which it has no use.

In a living organism these changes are not isolated
but are adjusted to one another so that the right
operations are carried out to produce the right
quantities of the various products. It is because

we are so impressed at the way in which all the separate processes work together harmoniously that we call animals "organisms." The processes which keep an animal alive have to be quite as highly organized as the operations in the most complicated mass-production factory. If there is a "secret of life," it is here we must look for it, among the causes which bring about the arrangement of innumerable separate processes into a single harmonious living organism.

When a numerous and varied set of processes is to be organized it is obviously convenient, and often absolutely necessary, to separate the different jobs among different pieces of apparatus, each of which specializes in carrying out one particular function. Thus a motor-car has a separate apparatus —the carburettor—to vaporize the fuel, another apparatus—the dynamo—to provide electric power, still another—the sparking plug—to make a spark, and so on. We find the same sort of plan adopted in all animals which attain more than a very minute size. For instance, every living creature has to arrange to absorb oxygen from its surroundings and to transport it in the right quantities to the cells in the body which need it. We find that there are special organs for absorbing it, lungs in animals which breathe air, gills in animals which absorb the oxygen dissolved in water; special organs, the blood-vessels, for transporting the oxygen all through the body after it has been absorbed and dissolved in the blood; a heart to pump the blood along; and many other organs to regulate the speed at which the lungs work and the blood flows. Without this rather complicated machinery, the organization of the oxygen supply would be inconceivable. The development of a set of specialized structures is the first step in the business of building up a living organism.

To say that an animal is an organism means in fact two things: firstly, that it is a system made up of separate parts, and secondly, that in order to describe fully how any one part works one has to refer either to the whole system or to the other parts. Thus it is impossible to describe fully a thighbone without referring to the fact that it is part of a leg, and that one end fits on to a pelvis and the other on to a shinbone. The relation with the other parts of the organism is indeed so close that if an anatomist finds a new fossil bone he can often reconstruct, in general outline, the whole unknown animal to which it belongs.

There are two possible ways of investigating the organization of an animal. Firstly, we can study in the adult how the organism works as a going concern: we can find out what functions are performed by each separate organ; we can discover how the communications between the organs are maintained by the blood and nerves; and we can study the results of removing one or more organs. But all the processes which can be investigated in this way will be proceeding within the framework provided by the fundamental spatial pattern in which the parts of the animal are arranged, since in the adult this pattern is more or less fixed. We can move large lumps of the pattern about, but we cannot discover what caused the pattern in the first place.

But there is a second line of attack. We can actually watch how the parts of a living organism come into being and fit together. Nearly all organisms start life as fertilized eggs, though a few grow out as buds from other organisms. Fertilized eggs are very simple-looking, often apparently quite homogeneous lumps of living matter. They consist of a watery jelly, the *protoplasm*, which contains a variable amount of food-matter or *yolk*, and which also encloses a little bag of special material which is the kernel or *nucleus*. As we shall see, the jelly-like protoplasm is not really as simple as it looks. But it is at any rate much simpler than the adult animal, which consists of very large numbers of cells, of several different kinds, arranged in various ways to build up the different organs. During the increase in complexity as the egg develops into the adult the spatial pattern of the animal arises. In the early stages it is fluid and unfixed; we can describe its gradual unfolding, make experiments which alter it, study its genesis and causation.

The study of development, or *embryology*, because it offers the possibility of finding out how the most fundamental characteristic of living things, their organization, comes into being, has always been of compelling interest to everyone who has been concerned with the position of living things in the general philosophical scheme. Nearly all biological philosophers, from Aristotle to the present day, have been embryologists. Aristotle, in fact, founded the science. He opened hens' eggs after they had been incubated for various lengths of time, and described what he saw. For centuries, embryology remained a purely descriptive science. The changes which embryos go through as they develop are so many and complicated that it took an enormous amount of careful and painstaking work simply to describe them. Scientists have always asked why the changes occur; but only in the last fifty years or so have they been able to perform experiments to try to find out; before that they could only guess, and, naturally enough, their guesses were usually wide of the mark. Even now we know very little about the causes which underlie embryonic development, but this is the most important and interesting part of the subject, and in this book I shall lay more emphasis on the tentative beginnings of our knowledge about the causes of development than on the description of the changes which occur.

One very important fact has been discovered and will be described later on in the book. It has been found that at a rather late stage the organization of an embryo is comparatively loose and the various parts are to a large extent independent of one another and of the whole embryo as regards the way they develop.[1] At an earlier stage, on the other hand, it

[1] Though not, of course, as regards the way they work: in this stage a lung can *develop* quite independently of the heart, but it cannot *function* to aerate the blood without the help of a heart.

has been found that the way any part develops is controlled in such a way that all the material which is available is worked up into one whole animal; and further, it has been shown that this integrating control is exerted by one particular part of the embryo. At this early stage, then, the embryo is very highly organized, because the way any part behaves in development cannot be described without referring to this special controlling part. The controlling part is therefore called the *Organizer*. I shall devote quite a considerable amount of space to a consideration of the organizers which have already been discovered, and the way in which they throw light on the other facts which have emerged in the study of development.

The main interest of embryology at present is theoretical, in the way discussed above. But there are a very large number of important practical questions which we may hope to be able to tackle later, when the science has been worked out more fully. For instance, why do most of the higher animals, including man, lose the power of regeneration so early in life, long before they are born? It would be very convenient if we could regenerate an amputated leg. Again, why do some cells start to form an unorganized cancerous growth which the animal cannot control, escaping from the agents which keep the parts of an organism together as one whole? How can we affect the production of twins from one egg? The answers to these questions are not, I think, right over the horizon of our present

view in embryology, but are quite near in front
of us.

The Similarity of all young Embryos

The study of embryology was given a great fillip by
the publication, and general acceptance among
scientists, of Darwin's theory of evolution. It had
already been found that the most general features
of an animal's organization, those by which it was
classified as a vertebrate, say, were formed early in
its development, and only later there arose the more
specialized characteristics by which it could be
classified as a bird or a mammal, while still later it
would develop the particular features of a fowl or
a duck. This means that in the very earliest stages
in development all embryos only show those characters which are common to all animals. They must
therefore look more or less alike. We have only to
describe the early stages and on the basis of their
common pattern we can make a general scheme
into which all embryos fit, and can classify the ways
in which they gradually diverge from each other.

Soon after the publication of Darwin's *Origin of
Species*, Haeckel put forward a general theory about
these early similarities. He supposed that each
animal, as it develops from the egg to the adult,
passes through a series of stages, each of which is
similar to one of its ancestors in the course of the
evolutionary history of the species to which it belongs.
The series is more or less in the right order, so that
the stage representing the first generalized ancestral

vertebrate occurs before the stages representing the various groups which gradually evolved out of the original vertebrate stock. This hypothesis brings under one head a large number of very odd facts. For instance, a young mammalian embryo, such as a young human embryo about four or five weeks old, is provided with gill slits and blood-vessels which flow along them. These are like the organs found in fish, where the blood flows through the gills and absorbs oxygen from the water, but they can be of no possible use to a mammalian embryo, which at this stage is deriving its oxygen from the blood-stream of its mother. Haeckel's theory, that such organs are "hang-overs" from the time when the ancestors of mammals were fish, still provides the most convenient way of describing this whole class of phenomena. But we have slightly modified the expression of Haeckel's theory. Many details of embryonic development are better described as reflections not of adult ancestors, as Haeckel thought, but rather of the embryonic stages of those ancestors. The mammalian embryo has gill slits, not like the gills its ancestors had when they were adult, but like the gill slits they had when they were embryos. With this modification, Haeckel's hypothesis, the so-called "biogenetic law" or recapitulation hypothesis, is still one of the foundations of our system of descriptive embryology.

But even so, there are very many features of development to which the law does not apply. Many embryonic characteristics do not represent any ancestral conditions, and not by any means all its ancestors are recapitulated in an animal's development. Embryos which live in special situations, like the bird embryo developing inside its shell, or the mammal in the womb of its mother, form peculiar organs suited to their particular conditions, and these often have little to do with any ancestral forms. We shall meet other examples of Haeckel's law later on, when discussing different types of larvae.

Haeckel's law is not strictly an explanation of anything. When a human embryo is developing, its remote fish-like ancestors are long dead and rotted to mud on the sea floor, and cannot possibly be the effective agents which cause the human to form embryonic gill slits. In order to give a satisfactory account of the direct causes of development, one must be able to show how the development is dependent on factors which are actually present in the fertilized egg or its immediate surroundings. What Haeckel did was to find a good way of *describing* the plethora of odd facts which had been accumulated. We ought to be able to find a reason why so many facts fit into Haeckel's generalization. Actually we still have not found any satisfactory reason, although we can suggest various processes which may be involved. Thus if there is to be an evolutionary change from fish into men, it is obviously easier, or so a man would think, to stick to the old plan of development until it is no longer a help and simply must be altered. The same thing happens in human inventions. When motor-cars were first made, the engineers did not think the whole problem out from the beginning and produce a stream-lined model with the engine over the back wheels where the power is required: they seem to have been exhausted by thinking out the engine, and simply attached it to the current form of horse-carriage in front where the horse had been. We can call this a sort of "habit reason"; men inventing, and embryos developing, tend to do what their fathers did if they can, because it is easier.

There may be other and more important reasons for imitating an old pattern. It may act as a guide. For instance, if one is going to make a cast-iron pot, one first models it in clay, as though iron had not yet been discovered; from the clay pot a mould is made into which the molten iron can be poured. Here the "ancestral" clay pot provides what may be called a formative stimulus for the "more highly evolved" iron pot. Perhaps this sort of explanation applies to ancestral characters which appear in embryos only for a short time, eventually disappearing entirely.

The Three Fundamental Layers

It follows from Haeckel's biogenetic law that young embryos must look much alike, since they should show only the characters which are common to all types of animals. This deduction from the law is actually true. Even before Haeckel definitely formulated his law quite a large number of different kinds of animals had been investigated, and it had been found that in their early development they all passed through the same two stages. These stages are called the *blastula* and *gastrula*, and presumably represent the ancestors from which all animals have been derived. The original form of Haeckel's law suggests that these ancestors looked like blastulae or gastrulae when they were adult, but no animals of this kind have survived till the present day. In fact, if we adopt the modification of Haeckel's law which was advanced above, there is no need to suppose that adult blastulae and gastrulae have ever existed; we need only assume that the original ancestral organisms from which all animals have been evolved passed through these two stages in their development.

There is quite a large amount of variation in the shapes assumed by the blastulae and gastrulae of different animals, but we can imagine ideal forms from which all the others can be derived by minor modifications. The ideal blastula consists of a hollow ball of cells, the walls of which are only cell-thick. The hollow inside is called the *blastula cavity*, or blastocoel. The ideal gastrula is also a hollow ball, but differs from the blastula in two ways; the ball is punctured, and the walls are thicker and consist first of two layers of cells and later of three. The hollow inside, together with the innermost layer which lines it, is the *primitive gut*, and communicates with the outside through the hole which punctures the ball. This hole is called the *blastopore*, because when it becomes visible as a little pore on the blastula surface it often provides the first visible indication that the blastula is changing into a gastrula. The three layers out of which the gastrula is made are named from the Greek words for skin, and for outside, inside, and middle; thus the outer layer is the *ectoderm*, the innermost layer the *endoderm*, and the layer between them *mesoderm*.

The ectoderm, endoderm, and mesoderm are the three fundamental parts out of which an animal is built. We might almost say that they correspond to the three major parts of a motor-car. The ectoderm develops into the skin, the sense organs, and the brain and nervous system; analogous to the body-work, the lamps, and the controls. The mesoderm forms the skeleton, muscles, and heart, or the chassis and engine. Finally the endoderm corresponds to the fuel system, and develops into the stomach and intestines and all the apparatus for absorbing food (i.e. fuel); this has to be much more complicated in an animal than the fuel system is in a car, because animals cannot get their nourishment poured into them in a form in which it can be used at once, as petrol is poured into the tank; it is as if a car had to carry round with it a whole refinery for turning crude oil into motor spirit. These three layers are called the *Germ-Layers*, and, when it was first propounded, the idea that they could be found in some form or other in all animals stimulated scientists to investigate as many different kinds of embryos as possible to see if the hypothesis was true. Most of this work was completed by the beginning of this century, and it provided a broad basis of information on which all our present-day knowledge of embryology has been built up. On the whole, it was found that the three germ-layers could be fairly easily recognized in most animals, but there are a few difficult cases, and in some very primitive animals only the ectoderm and endoderm are present and there is no mesoderm. The particular importance of the idea of the germ-layers from our point of view is that it is the beginning of an analysis of the pattern in which the embryo is organized. The formation of the three germ-layers is usually the first structural change which the embryo achieves, and almost immediately after this the main organs are formed. The process by which the blastula turns into the gastrula is known as *gastrulation*, and a great deal of the discussion later on in the book as to how development is brought about will be concerned with this period of gastrulation when the main structure is blocked out.

CHAPTER II

THE BEGINNING OF DEVELOPMENT

Egg and Sperm

THE history of a developing animal really begins when the egg-cell becomes fertilized by uniting with a sperm-cell, but before this can happen there must occur a very important series of processes by which

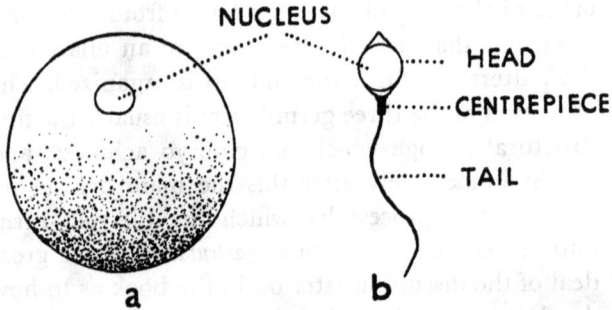

FIG. 1.—Diagram of an Egg-cell (*a*) and a Sperm-cell (*b*). The sperm is more highly magnified; in Man the egg is 0·2 mm. in diameter and the sperm about 0·05 mm. or 1/500th inch long, including the tail

these cells are elaborated. Egg- and sperm-cells are known collectively as *germ-cells*, and like other cells they consist of a mass of living matter or *protoplasm*, containing a *nucleus*. But in the details of their structure (Fig. 1) they are very specialized and unlike normal cells, as might be expected from the extraordinary things they have to do. The egg-cell is usually rather large as cells go, since it has to

contain food material for the embryo to use before
it develops a digestive apparatus. This food material
is stored as grains of yolk, which are scattered
through the *cytoplasm*, that is to say, all the protoplasm outside the nucleus. As it is fairly heavy, the
yolk collects at the bottom of the egg, which is
therefore stratified, with a yolk-laden *vegetative pole*
below and a non-yolky *animal pole* at the top. The
nucleus usually lies near the top, in the clear
protoplasm of the animal pole. Non-yolky eggs are
often not very much bigger than other cells; the
human egg, for instance, is about 0·2 mm. or a
hundredth of an inch in diameter. But when there
is much yolk, the egg-cell may be swollen to an
enormous size. The "yolks" of birds' eggs are single
cells, the biggest known, with only a tiny little patch
of cytoplasm nearly hidden in the huge mass of
yolk. The sperm-cell is still more highly specialized.
It consists of three parts: the head which contains
the nucleus, the centre-piece, and a long tail which
beats to and fro and drives the sperm actively about
through the fluid in which it exists. Sperm-cells are
very small, containing no yolk and hardly any cytoplasm, and their light construction enables them to
move about comparatively rapidly. A human sperm
can travel at the rate of about an inch in three minutes.
Eggs, on the contrary, are rarely able to move.

The most important part in the elaboration of the
germ-cells is the preparation of the nucleus, and this
process is essentially the same both in the eggs and
in the sperm. For most of the time the nucleus of

an ordinary cell consists of a bag made of the *nuclear membrane* filled with rather liquid protoplasm. When the cell is about to divide into two the nuclear membrane disappears, and out of the liquid contents there are built up a number of little solid lumps, which if the cell is killed can be stained very deeply with many dyes, and are therefore called *chromosomes*, from the Greek words for "colour" and "body." Different chromosomes are often different in shape, so that they can be recognized, and it is very important to notice that they always occur in pairs, so that each cell has two of each kind. The number and shape of the chromosomes in the cell is fixed for any particular species, but is different in different species; some have as few as four, others up to one or two hundred. But as the chromosomes are always in pairs of similar ones the number must always be even. When the chromosomes become visible at the beginning of an ordinary cell-division, each one is already split longitudinally into two half-chromosomes lying side by side. As the cell divides, these two halves separate from each other, and one half goes into each of the two cells which are formed. When the division is over they count as whole chromosomes, and gradually disappear into a normal fluid nucleus (Fig. 2).

The cell-division which results in the formation of the germ-cells seems superficially very different from the ordinary divisions, but it has recently been realized that the whole difference follows from one single slight alteration in the way the division begins. The difference is this: that when the chromosomes

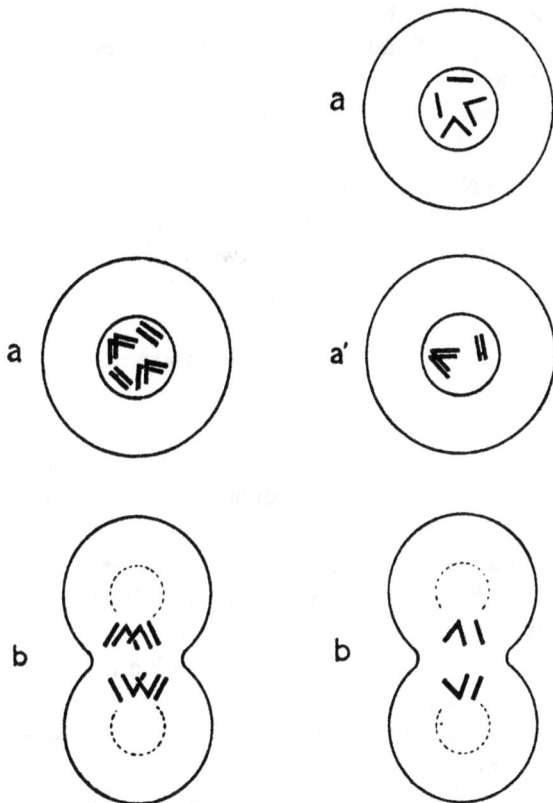

a

a

a′

b

b

FIG. 2.   FIG. 3.

FIG. 2.—Diagram of ordinary cell-division. (a) The chromosomes appear, already double, in the nucleus. (b) The cell
divides and one-half of each double chromosome goes to
each daughter cell

FIG. 3.—Diagram of cell-division during the formation of
the germ-cells. (a) The chromosomes appear single in the
nucleus. (a′) The two chromosomes of each kind lie side
by side. (b) The cell divides and one chromosome from each
pair goes into each daughter cell

appear before the germ-cell division they are not
split longitudinally (Fig. 3). They do not seem to
be able to proceed with the division until they have
arranged themselves into double bodies, to correspond with the two half-chromosomes lying side by
side which are found in ordinary division. They get
into a doubled condition in the only way which is
open to them; that is, by the two whole chromosomes
belonging to a pair joining up with each other and
lying side by side. If there are six chromosomes,
two A's, two B's, and two C's, for example, the two
A's always join up, and so do the two B's and the
two C's. The chromosomes are now a series of
paired bodies, which are just like the split chromosomes of an ordinary cell-division to look at, except
that there are only half as many of them. They go
on behaving just like the split chromosomes described
above; that is to say, the two partners in each paired
body, which have only just come together, now
proceed to separate, one partner going into each of
the two daughter-cells. This means that the two
daughter-cells have only got half the normal number
of chromosomes and are an exception to the general
rule in that they have only one chromosome of each
kind. That is one of the most important characteristics of the germ-cells. The ordinary body-cells,
which all have two of each kind of chromosome, are
said to have the *diploid* number, and the germ-cells,
which have only one of each kind, are said to have
the *haploid* number. The process which has just been
described is spoken of as the *reduction division* of the

chromosomes because it involves the reduction of their number to half.

The daughter-cells of the reduction division are not the actual germ-cells, but each one has to go

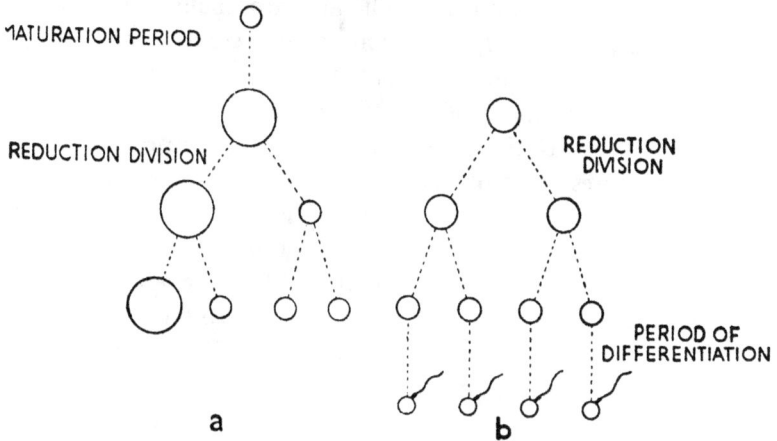

Fig. 4.—Diagram of the formation of the germ-cells. (a) The egg. (b) The sperm. The egg mother-cell is built up and furnished with yolk during the maturation period, and then undergoes two divisions, giving four cells of which three are very small and die. The sperm mother-cell divides twice and all four resulting cells are transformed into sperms during the period of differentiation.

through one ordinary division before it is ready, giving a total of four germ-cells from each cell which started the reduction process. Actually, in the formation of the eggs three of these four degenerate and never function as eggs, while the remaining one has to undergo a period of ripening when it is supplied with the yolk which it will require. This ripening usually happens in the middle of the reduction division, which therefore takes a very long time (Fig. 4).

Fertilization, Heredity, Virgin Birth

At the end of all this preparation the germ-cells are ready to carry out the complicated process of developing into an adult, and are finally ready for fertilization. Fertilization really consists of two processes, the activation of the egg by the sperm and the union of the egg and sperm nuclei. It is easy to see the importance of the second process; it restores the diploid number of the chromosomes by adding the haploid number in the sperm to the haploid number in the egg. A properly balanced set of chromosomes is essential for the development of the animal since they contain the hereditary factors.

An example will show how the influence of the hereditary factors can be detected. Men are sometimes born with short fingers, each with only two joints instead of three, because of some abnormality in the development of the fingers. The character is hereditary. For instance, we find short-fingered men who have married normal wives and all of whose children have short fingers. If two children born of such parents then marry, on an average one-quarter of their children will be normal and three-quarters short-fingered. These facts are due to the presence of a hereditary factor or *gene* for short fingers lying in a chromosome. The short-finger gene is an abnormal form of the gene which causes the fingers to develop in the ordinary way, and is derived from it by a sudden and as yet inexplicable change called a *mutation*. The original short-fingered fathers have

two similar chromosomes, each with a gene for
short fingers, and each of their germ-cells contains
one chromosome with the short-fingered gene. When
such a sperm fertilizes a normal egg containing a
gene for ordinary fingers, the children have one of
each kind of gene. In this particular case it is the
short-fingered gene which affects the development:
it is therefore said to be dominant over the ordinary

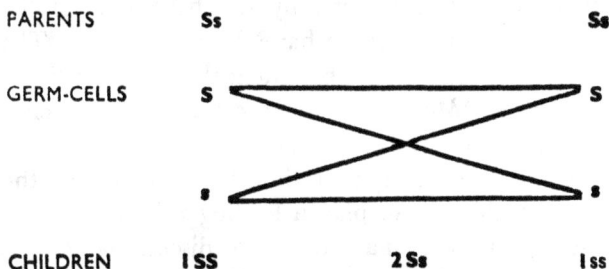

PARENTS   **Ss**   **Ss**

GERM-CELLS   **S**   **S**

                     s   **s**

CHILDREN   **I SS**   **2 Ss**   **I ss**

FIG. 5.—Diagram of the inheritance of short fingers. S is the factor
for short fingers and *s* that for ordinary fingers. The lines show
the ways in which the factors may come together in fertilization.

gene, which is recessive to it. When the germ-cells
are formed in the children of such a marriage, the
two genes, lying in the two similar chromosomes,
are separated at the reduction division, and the
germ-cells have half of them one normal gene and
half of them one short-finger gene. If two such
children marry (Fig. 5), it is pure chance which
genes come together in the fertilized eggs, so that
in half the eggs a normal gene will meet a short-
finger, giving short-fingered adults, in a quarter of
the eggs there will be two short-finger genes, giving
more short-fingered adults, and in the last quarter

there will be two normal genes giving normal adults. Hereditary factors of this kind were discovered by Mendel in the middle of the last century, and he also gave rules for the way in which they are inherited. Chromosomes had not been described at that time, and it is only about thirty years since it was realized that the hereditary factors actually lie on the chromosomes and that Mendel's laws are perfectly well explained by the behaviour of the chromosomes which we have described above. The theories propounded by Mendel are collectively known as "*Mendelism*" and are part of the science of *genetics*, or the study of heredity.

It is clear, then, that the chromosomes, or the genes within them, play a leading role in development, and we shall have to discuss later (see Chapter VII) how they do it. But we can say now that an egg can develop without a full double set of chromosomes; it can develop quite well, so long as it has got a half or haploid set. Anything less than this is fatal, and so usually is anything between a half and a whole set because of its lack of balance. Professor Dalcq in Brussels is investigating the development of frogs' eggs with less than the haploid number of chromosomes, and is finding out just when and how the embryos fail.

The other process in fertilization is the activation of the egg. We know very little about how this happens. What it does is to cause the egg to start dividing and developing. Now the same change can be brought about by other things which are not the sperm, and we then get a "virgin birth," or *partheno-genesis* as it is called in science. The most various and unexpected agents may be effective. It is sometimes only necessary to prick the egg with a sharp needle, or to put it into very weak acid; some marine eggs may be caused to begin developing if the salt-concentration of the water is altered. In all cases the procedures give rather variable results, and we have very little idea why they give any results at all. But the eggs treated in this way, since they have a haploid set of chromosomes, can go on developing quite normally. The adult which arises is smaller than normal, and its cells are smaller than normal, since they adjust their volume to that of the half-sized nuclei. Some eggs normally develop without being fertilized by sperm, i.e. parthenogenetically. This happens, for instance, to many of the eggs of bees, and these parthenogenetic eggs give rise to drones or males, which have only the haploid number of chromosomes. In some species such animals can produce sperm without performing another reduction division, but usually they are sterile. In other cases the egg starts developing parthenogenetically, and then succeeds in doubling its chromosome number, so that the diploid condition is restored.

Development Begins

The first steps in the development of the egg are always the same; the egg divides up into smaller and smaller cells without growing at all, till there is a mass of little cells in place of the large single egg-cell. This process is known as the *cleavage* of the egg. The details vary according to the amount of yolk which the egg contains. Eggs with very little yolk cleave into equal parts; those with rather more yolk

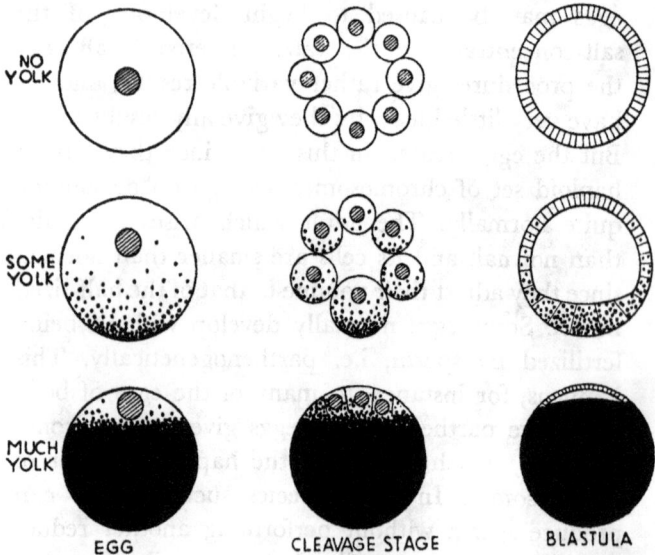

FIG. 6.—Diagrammatic sections of eggs, cleavage stages and blastulae with various amounts of yolk

cleave into small cells at the top or animal pole and larger cells at the bottom where the yolk is collected. Eggs with a great deal of yolk, such as birds' eggs, do not cleave throughout their whole mass : only the patch of non-yolky cytoplasm cleaves, forming a plate of little cells swimming on the surface of the main mass of uncleaved yolk. Such eggs are said to have partial or *discoidal* cleavage, as opposed to the *total* cleavage of the less yolky types. The total cleavage, as we have said, may be *equal* or *unequal* (Fig. 6), and further, it may be quite at random, forming a confused mass of cells; but often, particularly in the "total unequal" cleavages, it follows regular rules, so that the resulting cells are arranged in a definite pattern. An example of this, in the sea-urchin's egg, is described later (see p. 73).

The cleavage cuts up the large egg-cell into smaller cells, each of which, we can imagine, is more completely under the control of its nucleus than the unwieldy egg could be. It would be easy to suppose, and at one time it was supposed, that the nuclei divide unequally during the cleavage, so that the nuclei are unlike each other, and cause the cells in which they lie to develop in different ways into the various organs of the adult. But as a matter of fact this supposition is quite wrong: the cleavage nuclei are all alike. A very neat proof of this has been given by Spemann (Fig. 7). He tied a hair round a fertilized newt's egg, pinching it into a dumb-bell shape, so that the nucleus lay at one end and the other end had no nucleus. The end with the nucleus cleaved, while the other end did not. After several cleavages the knot was loosened and a nucleus, whichever happened by chance to lie nearest, allowed to pass through the bridge between the two ends. The second end now started to cleave, and it developed, not into any special part of the embryo depending on which nucleus it got, but into a whole embryo. In fact, the egg developed into twins. This experiment proved that any cleavage nucleus can develop into a whole embryo, and that

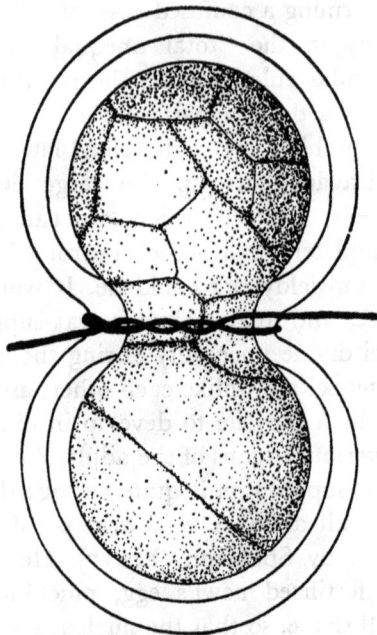

FIG. 7.—Spemann's tying-up experiment. (From Spemann.) The upper half has cleaved several times and a nucleus has just passed down into the bottom half, which has cleaved once

they are therefore all the same. Changes probably go on in the nucleus in later stages of development; but they are not important in the cleavage period, and do not determine how the parts of the embryo will develop. Even in later stages of development the genes still seem to be present in the nucleus and still active, since it is sometimes found that a gene has *mutated*, or changed into one of its other forms, in a body-cell of a late embryo, and it is then still capable of affecting the few daughter-cells which subsequently arise from it.

Spemann's experiment in which he made artificial twins raises the whole question of whether half an egg will always develop as well as a whole egg, and if so why an egg usually develops into only one adult and not two. But it will be more convenient to postpone discussing this problem (see Chapter vi) till we have dealt with the next period of development, the gastrulation period, in which the main outline of the embryo is formed. When the cleavages finish the egg is a blastula, a hollow ball of cells. The blastula-cavity appears quite early in eggs with a total cleavage, since the daughter-cells remain more or less spherical and leave a space in the middle, just as tennis balls would if they were packed together. This space grows until it is much larger than the individual cells, which continually decrease in size as the cleavages proceed. In large yolky eggs with a discoidal cleavage the hollowness is not so obvious, but the disc of cells lifts slightly away from the mass of yolk and leaves a narrow space which corresponds to the blastula-cavity (Fig. 6). The next chapter describes how this hollow ball of cells is converted into a three-layered gastrula.

CHAPTER III

MOVEMENTS AND FOLDINGS

In the blastula stage, the three fundamental layers
—the ectoderm, endoderm, and mesoderm—are all
simply different parts of the surface wall of a hollow
ball. If they are to arrive at their right places, with
the endoderm inside the ectoderm and the mesoderm
in between, a series of foldings has to be carried out.
There are several different methods of folding, as
we shall see, but they all lead to this same result.
We might compare the development of embryos to
making toys by folding up pieces of paper; but with
embryos, whatever the final shape, the folding always
starts by producing a three-layered gastrula, just as
if we are making toys we often start by making a
hat shape, and then go on to further foldings which
turn it into a boat or a frog or whatever it may be.

We shall have to describe some of the different
ways in which the gastrula is produced, both because
this is the most important process in the development of the embryo, and also as an example of how
the same process appears in a slightly different form
in different animals. In some embryos it is actually
a folding which occurs, but in others the layers
move into their right places by a streaming movement, sweeping across the surface and around the
inside of the gastrula like glaciers moving down a
mountain side. As a general rule, the lower the

animal is in the evolutionary scale, the simpler its gastrulation. The animals which are described here are selected partly because they have been particularly well investigated, and partly as typical representatives of this series of gradually increasing complexity. Thus we shall begin with the simple gastrulation of the sea-urchins and go on to primitive vertebrates like newts and lampreys, working up to more highly evolved groups of vertebrates like birds, and finally to mammals, which have developed a special container, the womb, in which development takes place.

The Formation of the Three Layers in Sea-Urchins

The simplest kind of gastrulation looks just as though one side of the blastula was being pushed and folded inwards by an invisible finger, in the same way in which one can push in one side of a rubber ball till it touches the other side. This happens, for example, in the embryos of sea-urchins and starfish. The first sign of gastrulation is a flattening of the bottom of the blastula, where the cells are usually slightly bigger than at the top, although this difference is not very striking, since there is not much yolk in the egg. Soon the flattened part sinks deeper in towards the centre of the blastula and makes a groove or small hole. This hole is a very important structural feature and turns up under all sorts of peculiar guises in other groups; it is called the *blastopore*, and is the entrance to the cavity lined by the endoderm, or the *primitive gut*. It grows deeper

and deeper as gastrulation goes on. The third layer
or mesoderm consists of a loose mass of cells, which
separate off from the endoderm (Fig. 8).

In Frogs and Newts

A rather more complicated gastrulation process
occurs in newts and frogs, which belong to the group

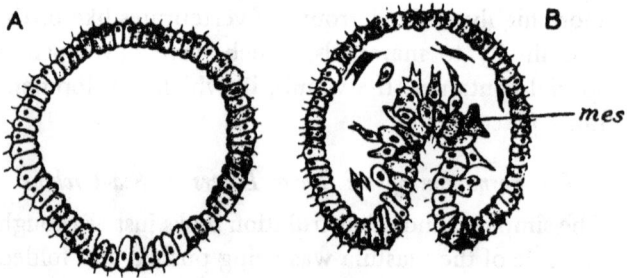

FIG. 8.—The blastula (*A*) and gastrula (*B*) of a sea-urchin. The
mesoderm is being formed from the endoderm in (*B*). (From
MacBride, after Field.)

of Amphibians. All amphibian eggs contain a fair
amount of yolk, which, as has been described, makes
the blastula asymmetrical and gastrulation rather
more difficult; moreover, being more highly evolved
creatures than sea-urchins, their gastrulation has
diverged more from the simple ideal type. In fact,
it is only by the application of modern methods of
investigation that we have found out what actually
happens. All the old methods involve killing the
embryos at different stages in their development and
comparing them. The embryos are usually killed in
such a way as to make them very hard, and then,

after embedding them in wax, they can be cut into thin slices or sections. These sections are then stained to bring out the different structures, and one can thus find out what the internal anatomy of the embryo is like. But for investigating the changes going on in gastrulation a very much improved method has been recently worked out, chiefly by Professor Vogt of Zurich. When gastrulation is beginning, the embryo is placed for a few minutes against little blobs of jelly which have been soaked in dyes. The cells of the embryo which are in contact with the jelly absorb the dye and become stained themselves, showing up as coloured patches on the surface of the blastula. If the dye is chosen rightly it is not poisonous, and the cells remain quite healthy and the embryo proceeds with its gastrulation. All one has to do in order to follow the gastrulation is to watch the dyed patches and see how they move. The whole process can be followed in one and the same embryo, and there is no need to rely on the comparative study of a whole set of specimens of different ages.

Fig. 9, p. 49, shows photos of a newt embryo during gastrulation. In the first photo gastrulation is just beginning, and a slight crescent-shaped groove has appeared on the surface of the blastula. This is the beginning of the blastopore. In the newt it does not lie right at the bottom, as it did in the sea-urchin, but rather to one side (see Fig. 10, a). From the outside, without the aid of coloured marks, all we can see is that the blastopore first becomes bigger,

and the horns of the crescent spread round till they
join up to form a circle, which encloses the whole
yolky bottom part of the egg; it then becomes
smaller again, pushing the yolky cells inside, and
closes up to a tiny slit, and eventually disappears.
Just about at the slit stage a sinuous ridge appears
on the surface of the embryo, marking off a horseshoe-
shaped area, which is darker in colour than the rest.
The sides rapidly get nearer together, and the
horseshoe-shaped area becomes squeezed up to a
dumb-bell shape, and finally to a deep groove. The
dark area is known as the *neural plate*, or *groove*,
according to the stage at which it has arrived. It is
the first rudiment of the brain and central nervous
system, and, as we shall find, gastrulation is completed before it appears; an embryo in which the
neural plate can be seen is no longer called a gastrula
but a *neurula*.

Fig. 10 shows a stain experiment from which one
can discover what is really going on all this time.
The first figure shows a series of patches made at
the beginning of gastrulation, arranged in a ring
round the blastopore. In the stages illustrated in the
next two figures the blastopore becomes round and
then closes up to a slit, and as it does so the patches
move in towards it and disappear inside, so that
when the neural plate has developed, in Fig. 10, *d*,
nearly all the coloured patches have been swallowed
up. A section through the embryo at this stage shows
that all three layers have been formed and that the
coloured patches lie in the middle layer or mesoderm.

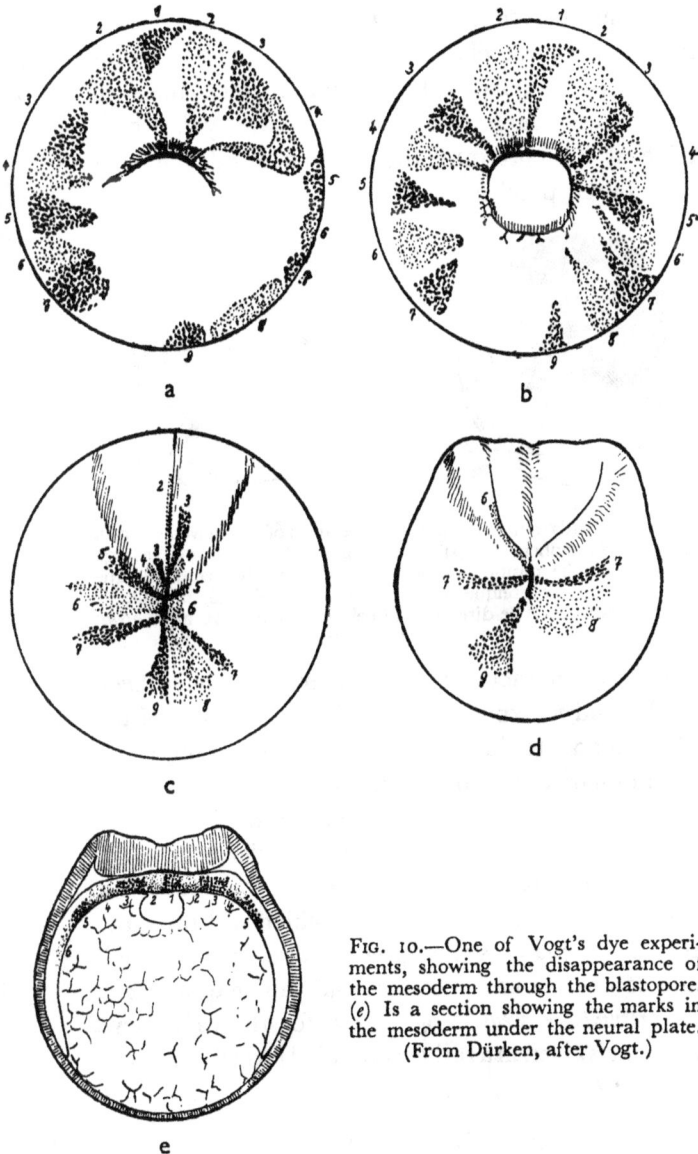

FIG. 10.—One of Vogt's dye experiments, showing the disappearance of the mesoderm through the blastopore. (*e*) Is a section showing the marks in the mesoderm under the neural plate. (From Dürken, after Vogt.)

They have moved nearer together and lie underneath
the thick neural plate. The primitive gut is represented by a narrow hole, and a thick mass of
endoderm, which was made from the yolky cells
lying at the bottom of the blastula, inside the ring
of coloured patches.

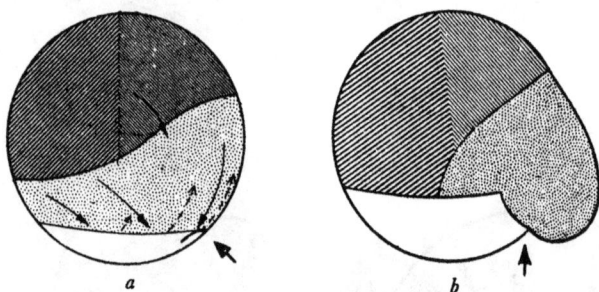

FIG. 11.—Vogt's map of the presumptive areas of the newt's gastrula
(a), and Weissenberg's map for the lamprey (b). Side view; the large
arrows mark the position of the blastopore. Widely spaced oblique
lines, skin; close oblique lines, neural plate; dotted, mesoderm;
white, endoderm. The directions of cell movements are also shown
on Vogt's map.

A large number of such experiments have been
made, and we know exactly where each part of the
blastula goes to during gastrulation. The easiest way
to summarize this information is to make a map,
showing for each region the organ into which it will
develop. It is usual to speak of mesoderm before it
has arrived at its final position as "presumptive
mesoderm" and of the material which will turn into
the neural plate as the "presumptive neural plate,"
and so on. We can therefore call this map a *map of
the presumptive areas*. Vogt's map for the early newt
gastrula is given in a rather simplified form in Fig. 11, *a*,

which also shows the directions in which the various regions have got to move to get to their final positions. The whole top of the blastula turns into ectoderm, the part nearest the blastopore into neural plate, the rest into skin. The presumptive endoderm lies at the bottom, and the presumptive mesoderm is an irregular ring between them, widest just in front of the blastopore. It is surprising to see the enormous movements which the tissues carry out. The presumptive skin has to expand so as to cover the whole surface; the presumptive neural plate has to swing in towards the middle line, while the presumptive mesoderm gets out of its way by diving into the blastopore and growing forwards along the inside. The endoderm is more passive, and is carried along by the other parts of the embryo. Nobody knows quite what makes these movements happen, or where the force comes from to push the tissues along. But already at the beginning of gastrulation the various regions have tendencies to behave in the appropriate way. If different bits of tissue are cut out and grafted back into other parts of the blastula (by a method which will be described more fully later), they go on behaving in their own typical fashion even in the wrong situation; the presumptive skin stretches in area, the presumptive mesoderm sinks in and disappears, and so on.

If we leave the coloured patches a bit longer, we can soon find out what the three layers develop into. We find that, as was said above, the ectoderm forms the neural plate, which turns into the brain and central nervous system, and also the skin. The mesoderm develops into the main muscles and skeleton of the body, and the endoderm forms the gut and all the organs which are derived from it later in development, such as the liver and lungs, etc.

Although gastrulation in the amphibian embryo is more complicated than in the sea-urchin, it is obvious that the two processes are similar in many ways. In both the net result is to finish up with three layers, and in both we find a blastopore and a primitive gut. But in the amphibia the mesoderm is formed directly from the skin of the blastula, and is not given off secondarily by the endoderm. Moreover, in the amphibian egg the process involves an active movement down to the edge of the blastopore, then a dive inside and movement again away from the blastopore along the inner surface, while in the sea-urchin the material simply folds in as a whole and there is not so much movement over the edge of the blastopore.

Lampreys

The "stain experiment" has not yet been performed on many groups of animals. Besides the amphibia worked out by Vogt, Weissenberg has done experiments with the lamprey, and Wetzel has given a full account of the gastrulation of the chicken. We need not give much description of the development of the lamprey, since it is extremely like the process we have just described in the newt. Fig. 11, *b*, shows

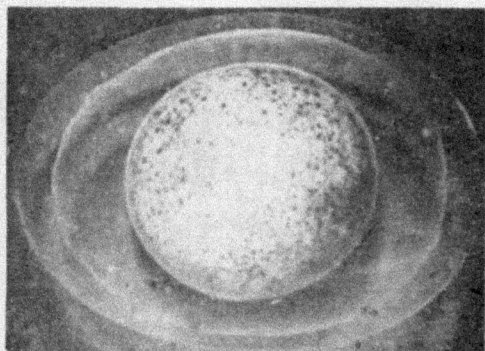

a

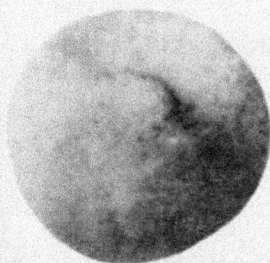

b

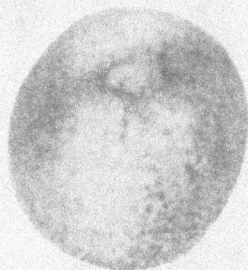

c

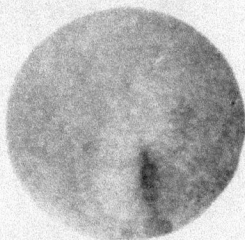

d

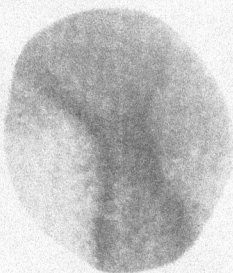

e

FIG. 9.—Development of the newt's egg. (a) The blastula in its membranes. (b) View from underneath of the early gastrula showing the beginning of the blastopore. (c) Later view of the round blastopore. (d) View from above of the slit-like blastopore at the end of gastrulation. (e) Later view showing the open neural plate

Weissenberg's map of the presumptive areas, and the similarity with Vogt's map is immediately obvious. It is an impressive example of the way in which different groups, here the fish and the amphibia, resemble one another in their early development. It should be mentioned, however, that the lampreys are a very primitive type of fish; the more highly evolved types have very yolky eggs and a different sort of gastrulation.

Birds

The chick embryo is a classical object for embryological investigations. The first studies we know of were made by Aristotle, and for a very long time no one looked more deeply into the matter than he had. The first sign of development which he could see was the appearance of a pulsating heart which is full of red blood and very large and obvious in the early stages. But gastrulation takes place still earlier, when the embryo is so small that the details can hardly be made out without a lens or a microscope. The process is, however, particularly interesting, because it is in some ways transitional to the conditions found in mammalian and human embryos. Both birds and mammals are evolved from extinct reptiles, and are therefore related to one another, though rather distantly.

The bird's egg, like the reptile's egg, contains a very large amount of yolk and undergoes the discoidal type of cleavage described in the last chapter, forming a blastula which consists of a little disc or cap of cells lying on top of the yolk but separated from it by a small cleft representing the blastula-cavity.

As in the sea-urchins, the gastrulation takes place in two stages, first the separation into ectoderm and endoderm and then the formation of the mesoderm. In birds the mesoderm is developed from the ecto-derm, not from the endoderm. The endoderm for-mation takes place very early, usually before the egg is laid. In the posterior part of the little disc, which is called the *blastoderm*, cells sink down and then turn underneath and grow forward along the under surface until the whole area is provided with a second layer, with ectoderm above and endoderm below. When the egg of a chicken is laid, the blasto-derm is already in this two-layered condition, and it is also divided into a central transparent area, where the embryo will develop, and an outer ring which is opaque because the endoderm cells there contain a great deal of yolk. The outer opaque area develops into various specialized organs for absorbing the yolk, and will not concern us any more in this chapter.

The changes which go on in the transparent area have been investigated by two new methods, and although their results are in almost complete agreement with one another, it is difficult to recon-cile some of their details with observations based on the old methods. One of the new methods is similar to Vogt's, and depends on watching what happens to coloured patches made with non-poisonous dyes; the other is to take cinematograph photos of the development. The cinema photography can be done in two ways: Gräper removed part of the shell over the embryo, stained the whole embryo with a non-poisonous dye so as to make it easier to see, and photographed it by reflected light: while Waddington and Canti took the whole embryo out

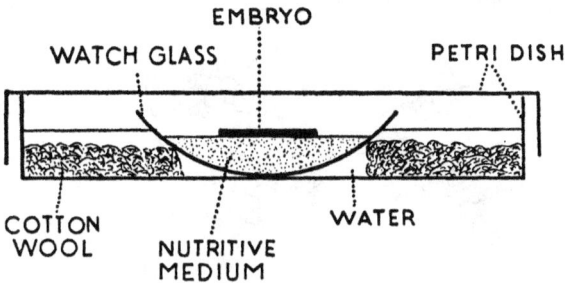

FIG. 12.—Glass culture vessel for growing embryos.

of the shell, cleaned the yolk off it, and put it on the surface of a nutritive jelly in a culture-vessel (Fig. 12), where it went on growing and developing and could be photographed as a transparent object.

When the egg is laid, then, the blastoderm has already developed for several hours after being fertilized in the body of the mother, and consists of a circle of transparent tissue surrounded by a ring of opaque tissue, both these areas being two-layered. The first thing that happens is that some of the ectoderm-cells collect together into a thickening which appears in the posterior part of the transparent area, and grows forward till it stretches as a narrow ridge across about three-quarters of the area. It takes about eighteen hours to grow to its full length, which is about 3 mm., so the rate of elongation is not excessively fast, only about a yard a month.

The mesoderm is produced from this ectodermal thickening or primitive streak, first in its front end and gradually further and further back. When mesoderm production begins the tissue at the sides

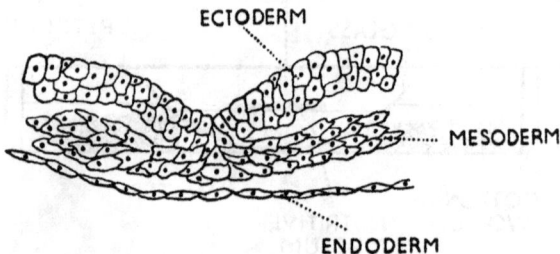

FIG. 13.—Section across the primitive streak of the chick showing the origin of the mesoderm

moves straight across the transparent area towards the primitive streak, dives down into it from each side, turns round, and grows out towards the sides again between the ectoderm and the endoderm (Fig. 13). If a little finely ground Indian ink is put on to the surface of the primitive streak and the embryo grown for a few hours in a culture vessel, sections will show how the Indian ink has been picked up by cells which originally lay on the surface and has been carried down into the mesoderm.

Finally, one more movement happens. The streaming in towards the primitive streak ceases, first in the front end, and instead a strong backward movement begins within the streak. The streak, in fact, gets shorter again, telescoping up along its length. The movement causes the front part to be drawn out, like glass in a flame, and as it becomes thinner and longer this part separates off from the rest of the streak and becomes the neural plate and the mesodermal rudiment of the backbone. This mesoderm does not remain as a flat plate, but soon breaks up on each side of the neural plate into a series of little lumps, the body-segments, which are found in all embryos at this stage, but are more obvious in the chick than in the other embryos which have been described. Between the two rows of segments, immediately beneath the neural plate, is a continuous strand of mesoderm, which is the first part of the backbone to be developed.

In a map of the presumptive areas at the stage when the primitive streak is fully grown but before the backward movement has begun, the presumptive neural plate and presumptive body segments and backbone are concentrated at the front end. The map in Fig. 14 shows this clearly for the neural plate. It is not very easy to compare this with the map for the newt, because in the newt the endoderm and mesoderm are formed together and in the chick are formed separately. The chicken has what may be called an "endoderm-blastopore" in the posterior of the blastoderm before the primitive streak appears, and Gräper thinks this should be compared with the newt blastopore; but Wetzel thinks the comparable structure is the later "mesoderm-blastopore"

in the front end of the primitive streak. The difference
of opinion shows how doubtful the question is: in
some ways one comparison is more informative, in
some ways the other. In both cases in front of the

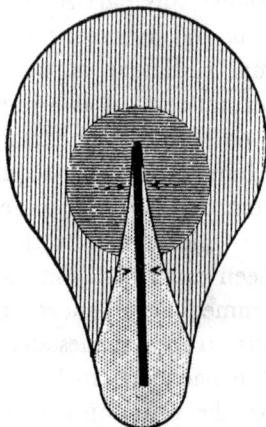

FIG. 14.—Map of the presumptive areas in the chick,
in the late primitive-streak stage, aged 18 hours. Only
the transparent area is shown. The arrows show the
directions of movement and the primitive streak is
drawn black. Widely spaced vertical lines, skin; close
horizontal lines, neural plate; dotted, mesoderm. The
endoderm lies underneath the whole surface

blastopore there is a zone of presumptive mesoderm
and outside that a zone of presumptive neural plate,
just as there is in the newt or lamprey. In the map of
the later stage (Fig. 14) part of the mesoderm has
already disappeared inside.

Mammals

We must now go on to consider the development of
mammals, but it will only be possible to give a very

rough sketch of their gastrulation in the space available, because different members of the group have evolved quite different ways of developing. Probably the extreme diversity of the processes of gastrulation in mammals is due to the difficulties raised by the fact, so advantageous in many ways, that the embryo remains in the maternal womb. There are actually two main problems which arise. In the first place, although the mammals are derived from reptiles whose eggs contain a large quantity of yolk, they themselves produce eggs which are extremely poorly provided with yolk, all the necessary nutriment being derived from the mother. Secondly, a whole set of organs have to be developed to make an adequate connection with the wall of the womb. The mammal, having evolved through a stage with yolky eggs, recapitulates this stage in its development, in accordance with Haeckel's recapitulation hypothesis, but it converts the organs which in its ancestors were used to absorb the yolk into the placenta and other organs by which it is attached to the mother. The placenta is porous and allows oxygen and food substances to diffuse from the maternal blood into the embryonic blood, which is carried along the blood-vessels in the umbilical cord into the foetus, which is thereby nourished and sustained.

The mammalian embryo, again like its ancestors the reptiles, goes through a primitive streak stage, but as the egg contains very little yolk it arrives at this stage in rather a different way. The cleavages

are quite normal: they are total and nearly equal
A blastula cavity appears in the usual way, and
the cavity grows to an enormous size (Fig. 15). The
walls are very thin, only one cell thick, except at
one place where there is a little lump of cells hanging
down inside like a drop or a swarm of bees. This
lump is the *inner cell mass* and eventually develops
into the embryo, while all the rest of the thin-walled
blastula develops into placenta, etc., and corresponds
to the opaque area in the chicken's egg.

The endoderm forms by a process which is so
simple that it can hardly be called a gastrulation;
the lowest layer of the inner cell mass simply grows
out in all directions till it covers the inner surface
of the blastula. Sometimes the mesoderm, or at least
part of it, forms in a similar way by differentiating
in situ from the cells of the middle part of the inner
cell mass. But some mesoderm is also formed as it is
in chicken. That is to say, it comes from the primitive
streak, which appears on the top of the inner cell
mass, in the upper layer of cells, which is the
ectoderm. Part of the inner cell mass may also split
off to form more of the non-embryonic apparatus,
but the embryo always passes through a stage when
there is a thickened area, the upper layer of which is
the ectoderm with the primitive streak, the lowest
layer the endoderm, with probably some mesoderm
between. It is not known exactly how much mesoderm is formed by the primitive streak, and no
staining experiments have been done, so it is impossible to give anything like an accurate map of

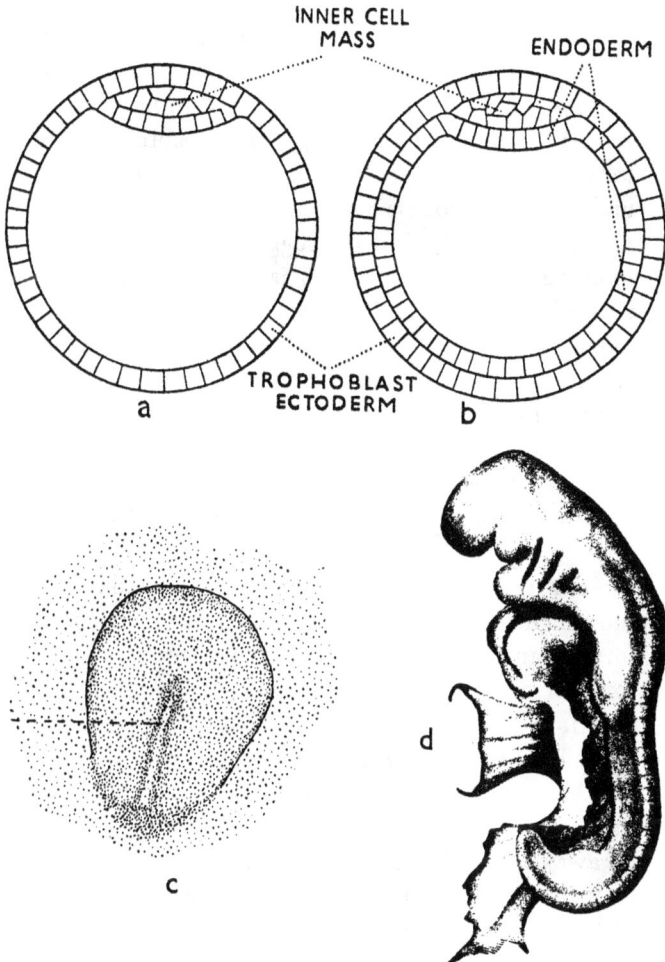

FIG. 15.—The development of the rabbit. (a) and (b) are sections through the whole egg showing the formation of the endoderm, (c) is a surface view of the primitive streak which develops on the inner cell mass, and (d) (from Quain's *Anatomy*) is a much later stage showing the gill slits

the presumptive areas. It may be possible to work out such a map in the near future by growing the embryos in cultures like the chicken embryos. It is already quite possible to keep rabbit embryos alive for a few days, so that they develop fairly normally, and it should not be very difficult to improve this technique slightly. But the suggestion that man will soon be born out of bottles is very optimistic (or pessimistic, according to the point of view). The technique of cultivating embryos will probably remain a useful scientific method for a long time, but it is difficult to imagine it being used practically.

The mammalian embryo, including the human embryo, therefore goes through a primitive streak stage very like a chicken. Although we do not yet know much about the details of what is happening, there are various reasons for believing that the movements which take place are similar in both sorts of animals. We can draw this conclusion from some of the kinds of abnormal embryos which are found. Gastrulation is a process involving such complicated movements that it is quite easily upset. For instance, the material which moves forward to build up the primitive streak may for some reason go awry and split into two separate streams, so that it makes two front ends of the streak, and the embryo develops with two heads. Or a similar splitting may happen later, when the presumptive neural plate and body-segments are being pulled backwards into their final positions, and we then get two posterior ends in the embryo, which is split part way up the back. Both these sorts of monstrosities sometimes occur in man, if anything happens to disturb development, but the embryo is so well guarded from casual disturbances, well hidden as it is in the womb, that they are fortunately rare.

It is interesting to notice that mammalian embryos are relatively very slow in arriving at the primitive streak stage of development, but then run ahead quickly till the main parts of the body are present, slowing down again, in the case of man at least, for the final elaboration of the parts before birth. Exact comparisons of the number of days taken by different embryos to arrive at comparable stages are not very informative; what is important is not the absolute times but the relative times for various amounts of development within the same life-history. Thus a newt takes about one day to get to the beginning of gastrulation, about one day to gastrulate, and about two days to develop from the end of gastrulation to an embryo with fifteen body-segments (the times actually depend on the temperature, since the warmer the egg is kept the faster it develops; but the temperature does not seriously affect the relation between these lengths of time). A rabbit embryo takes nearly eight days to get to the beginning of gastrulation, and then like the newt about one day to gastrulate and two to develop into an embryo with fifteen body-segments. A human embryo takes about a fortnight before arriving at the primitive streak stage. This slowness in early stages, and speed in later stages, is probably due to the fact that development within a womb is a fairly new idea in evolution; mammalian embryos are not yet perfectly adjusted so as to take advantage immediately of the favourable condition which a womb provides.

The length of time between conception and birth is different in different mammals. On the whole, the bigger the animal the longer the time, but there is not a strict proportionality; the smaller ones take longer than would be expected, and so does man. In elephants the time is about 620 days, during which time new elephant tissue is formed at an average rate of 14 lbs. per day, while the mouse, which is only one-quarter-millionth as big, takes 21 days to produce an infant, making new "mouse" at the rate of only a fiftieth of an ounce per day. Bigger animals also live longer than smaller ones, and the whole tempo of their lives is slacker. Perhaps each animal has its own apprehension of time, so that for instance a mouse feels a minute of man-time as a whole quarter-of-an-hour of mouse-time, while to the elephant it is only a few seconds of elephant-time.

CHAPTER IV

THE "ORGANIZATION CENTRE"

IF we observe the development of an egg, it is fairly easy, as we have shown in the previous chapters, to find out what each part develops into, but much more difficult to discover why it develops as it does. But the solution of this problem is fundamental for an understanding of living organisms. It was pointed out in the first chapter that the functioning of living animals depends on their structure and, since they are not man-made machines, but are such that the ability to develop is an essential part of their nature, we have to be able to give an account of how this structure arises before we can understand them.

One cannot expect to answer at once the question of why an egg develops at all. It must be unstable in some way which we do not understand, and bound to begin changing and developing as soon as it is fertilized. But it is better to begin investigating something simpler; let us ask, why does one part of an egg develop into one organ and another part into another? The factor which determines what a given part of the egg will become, or determines its *developmental fate*, as it is called, might be either inside or outside the piece of material in question. That is to say, it may be that a given piece of the egg will develop in accordance with a pre-determined developmental fate under any conditions in which it can develop at all, or, on the other hand, the development of the piece might be directed by something outside it. Both possibilities are actually realized. In the first case the piece is said to be *determined* and to be capable of *self-differentiation*. In the second case the tissue is undetermined, and the external factor which determines it may be either in the other parts of the egg or else outside the egg altogether. It is unlikely that all the determiners will be outside the egg, since it is difficult to suppose that, if an egg is floating in the sea, for instance, the sea water surrounding its different parts would be sufficiently dissimilar to cause the development of a whole set of organs. Actually the external determiners are mainly important in very early development though they are active again to cause minor modifications towards the end of embryonic life. The first difference between the various parts of an egg *must* in any case be externally determined, either during the process by which the egg is formed in the body of the mother or soon afterwards; if the egg is originally made with all its parts exactly alike it cannot on its own develop special characteristics in its different parts because there is no reason why a special character should arise in one part rather than another. Actually it is found that all eggs are made with certain differences between the parts; there is, for instance, an animal pole at the top and a vegetative pole where the yolk is collected at the bottom. Often the entry of the sperm at a definite point is a further important external determining factor. But the number of these early external determiners is comparatively small, and the most striking steps in development are caused by internal determiners.

The Technique of Micro-Surgery

The first experiments which were successful in locating these determiners inside the embryo were carried out by Spemann. He used two species of newts, one of which, *Triton cristatus*, lays a white egg, while the other, *Triton taeniatus*, lays a rather dark-coloured egg. Spemann first had to work out a technique for making operations on these eggs. The eggs themselves are embedded in two layers of jelly, but these can be removed fairly easily. However, when the eggs have been made free in this way, there are further difficulties because the eggs are so small, only about 2 mm., or a tenth of an inch, across. Spemann invented two special instruments for cutting such small masses of tissue; one is made of very fine glass drawn out to a sharp point, much finer than any point which can be made on a pin, and the other is a little loop of baby's hair, mounted on a holder, which can be used as a knife. All the operations have, of course, to be performed under a microscope.

The advance of experimental embryology is always waiting on two things, the invention of new instruments and methods, and the training of people with sufficient manipulative skill to use them. Spemann's instruments were the first which were invented for the purpose of making very minute operations on embryos. They are not difficult to use moderately well, but it requires years of practice to become really skilful with them. For working on chick embryos, which will be discussed on page 70 f, Spemann's instruments are useless because the tissue is too tough to be cut with the fine glass points; one has to use coarser steel knives, which are rather more difficult to handle. No one has yet invented instruments really suitable for working on mammalian embryos, and that is partly why we know so little about them. The tissue is so sticky that it clings to the end of a knife, and if one tries to scrape it off, it usually gets torn to bits and spoilt; this small technical difficulty makes the most important experiments impossible to perform.

The Focus around which the Embryo is Integrated

Speman's first experiment was to make an exchange between a piece of white presumptive skin of *cristatus* and a piece of brown presumptive neural plate of *taeniatus*. He did the same operation at various stages of development, and found that if the exchange was made between two early gastrulae the result was quite unlike what happened if it was made between late gastrulae. When the experiment was done in the early stage, the piece of white *cristatus* presumptive skin which had been grafted into the neural plate region of the *taeniatus* egg, developed into neural plate like its surroundings; and the brown *taeniatus* presumptive neural plate,

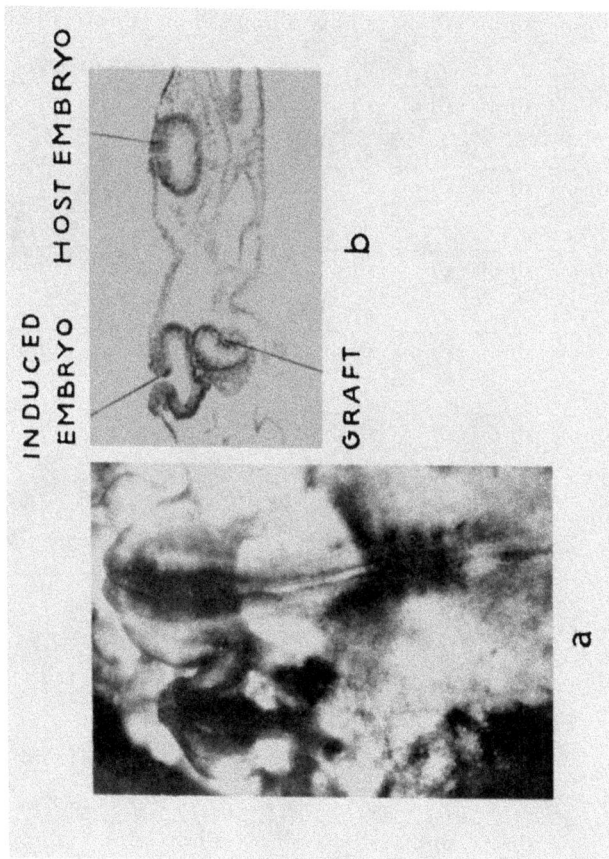

INDUCED HOST EMBRYO
EMBRYO

GRAFT

a

b

FIG. 18.—An organizer graft in the chick. (a) Shows the host embryo with the induced head just to the left of it. (b) Is a section

having been grafted into the *cristatus* skin region, developed into skin (Fig. 16). Both the pieces of tissue,

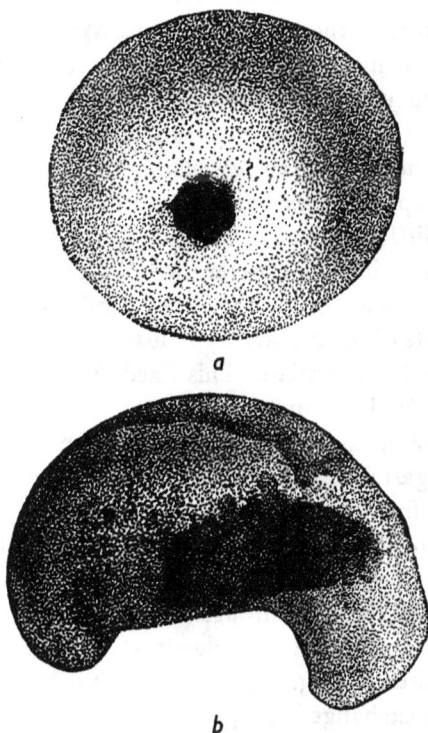

a

b

Fig. 16.—Spemann's exchange experiment (from Dürken, after Spemann). (*a*) A gastrula of *Triton alpestris* with a small dark implant of the presumptive neural tissue from *Triton taeniatus*. (*b*) The *alpestris* embryo at a later stage with the *taeniatus* tissue forming part of its skin

in fact, developed in accordance with the region into which they were grafted, and not in accordance with the region from which they originally came.

They did not self-differentiate and therefore cannot have been determined when the experiment was made: their fates must have been determined by something in their new situations. When the experiment was made in the late gastrula stage, the result was the exact opposite. The white piece of presumptive skin developed as skin although it was lying in the middle of the *taeniatus* brain, and the transplanted piece of *taeniatus* presumptive brain developed into a little patch of brain lying in the middle of the *cristatus* skin. The determiner, then, must have finished its action by the late gastrula stage, so that the parts of the egg are by that time determined and must self-differentiate. This fixed the period during which the determiner works.

Spemann then began to examine the various parts of the gastrula, trying to locate the determiner. He got his first hint of where to look from the following experiment. He took a young *taeniatus* gastrula, cut it in half with a horizontal cut, and then turned round the top half through half a turn and replaced it in its new position. A glance at Vogt's map of the newt gastrula (Fig. 11) will show what he had done; he had exchanged the presumptive skin and neural plate areas. However, the gastrula healed up again and went on developing. The neural plate appeared in front of the blastopore as usual; in fact, in its normal position as regards the bottom half of the egg. But it was in an abnormal position as regards the top half and must have been made out of presumptive skin. This shows that the determiner must be in the bottom half; Spemann suggested that it is actually located near the blastopore.

A few years later Spemann and his pupil, Hilde Mangold, tested this suggestion. They removed the blastopore region from a young *cristatus* gastrula and grafted it into the belly region of a *taeniatus* gastrula. The *taeniatus* gastrula, which thus had two blastopores, developed two neural plates, and two complete sets of embryonic organs (Fig. 17). This might have been due simply to a self-differentiation of the grafted blastopore, but Spemann and Mangold showed that this was not a complete explanation. Owing to the differences in colour between the cells of the *taeniatus* host and the *cristatus* graft, they could tell how the secondary embryo was built up. Examination showed that the grafted blastopore had developed chiefly into mesoderm, which was its presumptive fate, while the neural part of the secondary embryo had been formed from the tissues of the host. The mesodermal part therefore *had* arisen by self-differentiation of the grafted blastopore, but the neural part had *not*. It had been *induced* out of host tissue which would ordinarily have formed skin. The implanted blastopore had therefore determined the fate of this presumptive skin, causing it to become actual neural plate. This experiment definitely locates the determiner at the blastopore. It was later shown that the whole ring of mesoderm which lies in front of the blastopore in Vogt's map is capable of determining, though this capability is strongest just near the blastopore and becomes

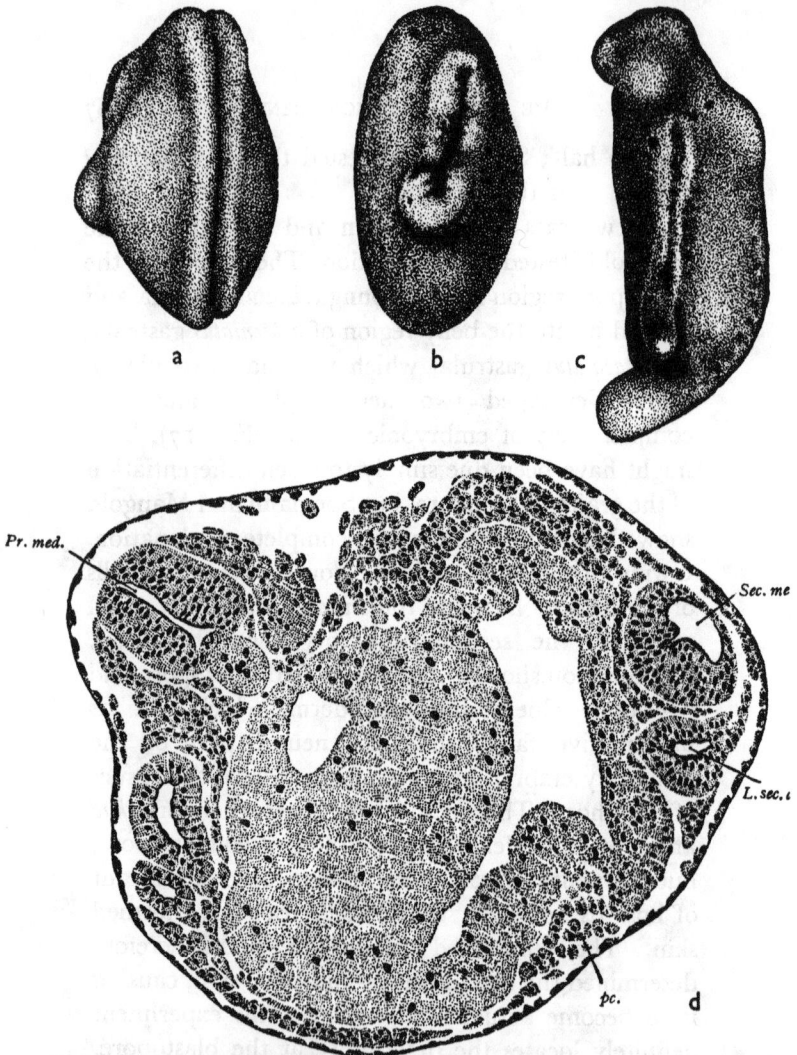

FIG. 17.—An organizer graft in the newt (from Dürken, after Spemann and Mangold). (a) Shows the neural groove of the host embryo with the induced neural groove on the left. (b) Shows the induced neural groove more clearly. (c) Is a later stage, the host's neural tube is on the right and the induced neural tube runs straight up the middle. (d) Is section with the host's neural tube on the left (pr. med.) and the induced neural tube (sec. med.) on the right, with an induced ear (l. sec. aud.) beside it

weaker further away from it. Spemann gave the name of *organizer* to pieces of tissue which can determine in this way, and of *organization centre* to the part of the embryo where they occur, but these two names are often used more or less interchangeably in a rather loose way. The organization centre is the whole presumptive mesoderm, which, as we know, sinks in through the blastopore and grows forward along the inside. Finally the mesoderm spreads over the entire embryo between the ectoderm and the endoderm, but at first it forms a narrow tongue which is the roof of the primitive gut, and it is in this stage that it determines the ectoderm lying immediately above it to become neural plate, while all the rest becomes skin.

The importance of the organization centre can be exhibited in two ways. One way is to point out that it causes part of the gastrula-ectoderm to develop into neural plate. This property is important enough, but other similar development-provoking agents, or determiners, had previously been discovered. The special importance of the organization centre is better conveyed by the name Spemann actually chose; it is that part of the embryo with respect to which all the rest is organized. In order to describe the behaviour of any part of a newt gastrula, it is necessary and sufficient to specify its relation to the organization centre. Spemann's name for his discovery may at first sight seem rather grandiloquent, but is really quite reasonable and accurate.

Birds

This discovery is so fundamental that we must go on to see if it applies to other animals and try to find out if they also have organization centres. Work which has been carried out since Spemann's original experiments has shown that in fact organization centres are not merely a speciality of the newt's egg, of no general importance, but, on the other hand, can be found in nearly all groups of vertebrates and in some invertebrates. The one which is perhaps most similar to the amphibian organization centre has been found in the birds, but, as we might expect, there are differences correlated with the differences in gastrulation in the two groups.

As was described in the last chapter, the endoderm and mesoderm are formed separately in the bird, while in the newt they are both formed at the same time as the blastopore. We find that not only are the formative functions of the newt's blastopore shared between two structures in the bird but so are its organizing functions. There are, in fact, two bird organizers instead of one; the first is concerned with the formation of the endoderm, the second with that of the mesoderm.

All the experiments with birds were done on chicken or duck embryos which were kept in culture after they had been operated on. The first experiment was concerned with the endoderm. This is already formed when the egg is laid, so there is no possibility of transplanting the "endoderm-blastopore." Instead, young embryos were taken at a stage when the primitive streak or thickening had just appeared; the endoderm was removed, and then replaced after being turned round through 180 degrees, so that the part which originally lay under the primitive streak in the posterior region near the endoderm-blastopore now lay under the other, anterior end. In some of these operated embryos an extra primitive streak appeared in the anterior end, where it had been induced by the endoderm coming from the endoderm-blastopore region, which is therefore an organization centre. Meanwhile the original primitive streak, which had been induced by the endoderm before the operation was made, continued to develop, so that finally two embryos appeared on the one blastoderm.

The next experiment dealt with the formation of the mesoderm from the primitive streak at a rather later stage. It is easy to take a piece of primitive streak from one embryo and place it between the ectoderm and endoderm of another; it develops by self-differentiation in its new situation into its presumptive fate, which is mesoderm and a little neural plate tissue. But the important thing is that the host ectoderm lying above it, which ought to turn into skin or part of a yolk-absorbing structure, now is induced to form a secondary embryo with a neural plate and other organs. This proves that at least some of the presumptive mesoderm, that part which lies at the front end of the primitive streak, is also an organization centre (Fig. 18), as it is in the newt embryo.

Mammals

Organizers have been found in other groups still more widely removed from newts. We know rather little about the organizers in mammals: in man we have no certain knowledge at all of them. But it has been proved that the rabbit embryo can react to a chick primitive streak organizer and can be induced by it to develop an extra neural plate. This suggests that the mechanism is essentially the same in mammals as in the other groups and that they also have organizers, but it has not yet been possible to make organizer grafts because of the technical difficulties mentioned before (p. 64). This experiment also shows how extremely unspecific organizers are. It seems that an organizer from any group of animals will work on the embryos of any other group. Probably, as is discussed later (p. 94), the activity of the organizer is due to a chemical substance, and this may very likely be the same substance in all the different groups of vertebrates.

Sea-urchins

Among the invertebrates, Hörstadius has discovered an organization centre in the sea-urchin embryo which is very like, though slightly different from, that of the higher animals. The sea-urchin's egg, as was described earlier, has a very simple gastrulation; the blastula is spherical, with all the cells more or less equal in size, and in gastrulation the bottom of it just draws back into the inside. Gastrulation occurs quite soon after fertilization, when the cleavage cells are still rather large; and if one attempts to make grafts such a large proportion of cells get injured in the operation that the experiment is not a success. So Hörstadius did his operations still earlier, during the cleavage, when he could cut in between the cells without injuring them. Most of the experiments were made when the egg had divided into sixteen cells. The cleavages are not quite equal, and the sixteen cells are arranged in three circles, one at the top consisting of eight moderate-sized cells, then a circle of four large cells, and at the bottom a circle of four small cells (Fig. 19). Hörstadius cut the embryo in half in various ways and then joined the halves together again. He found that several, but not all, combinations of halves could regulate themselves so as to produce a normal larva. He also found that wherever the small vegetative cells were grafted, they started sinking in and turning into endoderm, and moreover they carried the surrounding cells in with them and induced them to become endoderm. They therefore acted to some extent like an organizer. We shall discuss in Chapter vi the way in which this organization centre seems to differ from that of the newt.

Insects

In the insects another sort of organization centre has been found which is still more different from that of the newt. It was discovered by Seidel, who worked on dragon-flies' eggs. We have not said anything

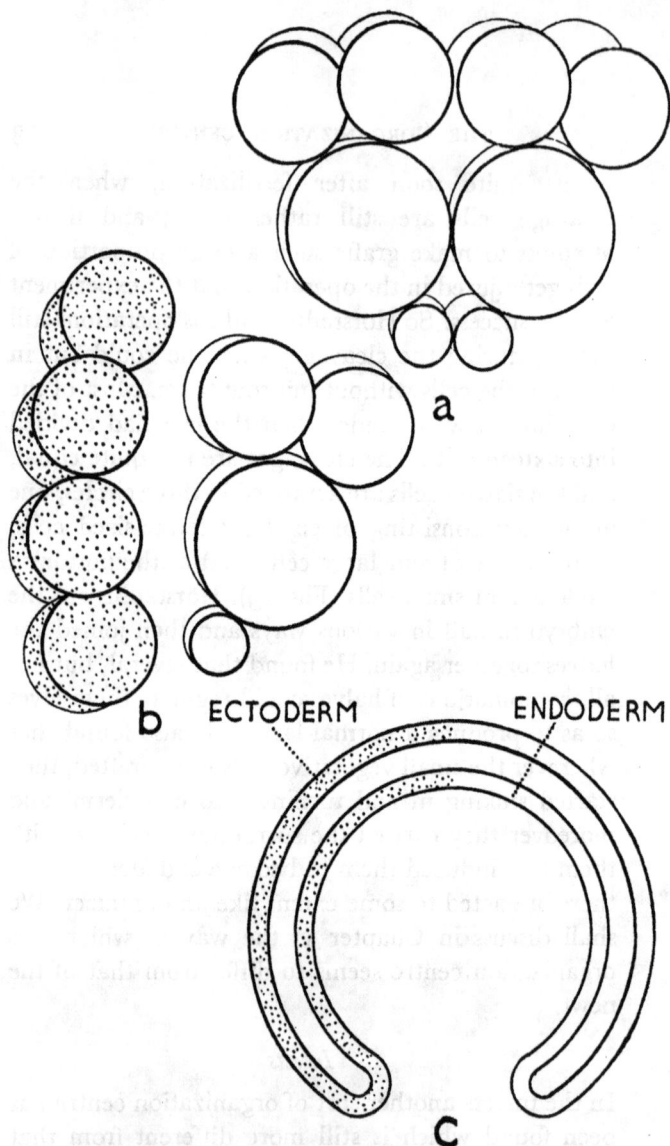

FIG. 19.—Grafting experiment in sea-urchins. (a) Shows the normal 16-cell stage; in (b) a top half of one egg (dotted) is combined with a side half of another. (c) Is a section of the resulting blastula showing some of the dotted material turned into endoderm

about the development of insects yet, and it is rather unlike the development of most other animals. The egg is a banana-shaped object, containing a large amount of yolk. The nucleus, after it is fertilized, lies nearly in the middle. The "cleavages" consist of divisions of the nucleus without any accompanying divisions of the whole mass of yolk, so we find gradually more and more nuclei scattered about throughout the egg, till the whole yolk is full of them. After a time the nuclei collect on the surface and cell boundaries form round them, so that eventually a stage is reached with the yolk nearly empty of nuclei and covered with a cellular skin. This is called the blastoderm stage and corresponds to the blastula, only here the blastula cavity is filled up with yolk. Then gastrulation takes place, starting with a sinking in of the blastoderm-skin at one side of the embryo (Fig. 20).

Seidel tied thin hairs round the egg in the early "cleavage" stages, and thus constricted the egg into two parts. He then watched how each part developed and in this way could discover the influences each part had on the formation of the embryo. He found that there is an essential region in the bottom end of the egg. He called this an *activation centre*, and it is not quite the same as an organization centre, because it does not determine the way in which the various parts of the egg will develop. What it does do is to activate another centre which lies further towards the front of the egg. This other centre, which is called the *differentiation centre*, lies, in fact, just where the embryo begins to be formed by the thickening and sinking-in of the blastoderm. One might almost say that it lies at the blastopore, like the other organization centres, but the gastrulation of the insects is so a-typical that such a comparison is doubtful. In any case, the differentiation centre is

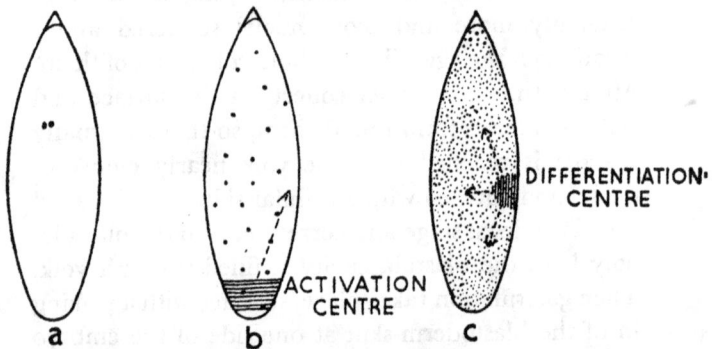

FIG. 20.—Diagram of the development of the dragon-fly. (a) Shows the egg- and sperm-nuclei just about to unite. (b) Is an early cleavage stage with the activation centre. (c) Is the "blastula" stage, with many nuclei lying on the surface, when the differentiation centre is active

much more like an organization centre in its activity than is the activation centre. As soon as it has been activated it starts to form a complete embryo round itself. If part of the egg is tied off with a hair, so that only a small region remains connected with the differentiation centre, nevertheless a complete embryo may be developed in the part. This means that the centre alters the fate of the tissues around it and determines how they shall develop so as to give a complete embryo. Seidel discovered how at least the first step of this determination is done. The differentiation centre starts a wave of contraction passing through the yolky core of the egg, and as the yolk contracts away from the outer skin of the egg, a hollow is left in which the blastoderm cells rapidly collect and begin to develop into the embryo. If part of the egg is tied off with a hair, this hinders the progress of the wave of contraction, which adjusts itself to the amount of space which is available to it.

Seidel also found out something more about the activation centre. He showed that it could only be formed if the early cleavage nuclei are allowed to wander down to the bottom end of the egg, where they react in some way with the cytoplasm there and set free a substance which diffuses forward again through the yolk to the differentiation centre. Neither the nuclei alone, nor the particular region of cytoplasm, can act effectively.

There is one very important respect in which all the organization centres we have described are alike, except the Activation Centre, which is rather in a class apart. Not only do they all become active when gastrulation starts, but they are all located at the centre of the gastrulation movements, that is to say, at the region of the blastopore. This is true both when it is endoderm which is being formed, as in the sea-urchin or the first chick organizer, and when the important process is the organization of the mesoderm, as in the newt or the second chick organizer. The blastopore is not only the centre of the movements which are going on but also of the formative influences which are at work.

CHAPTER V

THE ADDITION OF DETAILS

Secondary Organization Centres

AFTER the organization centre has acted in the newt or chicken embryo, the main outlines of the future animal are laid down. But the embryo is still very undeveloped and the outlines have to be filled in with all the multifarious details of the adult; none of the minor organs, such as eyes, ears, lungs, and legs have yet appeared. Many of these minor organs are also developed as responses to organizing stimuli. The organization centres which have been described so far are, in fact, only the primary organization centres, and they are succeeded by secondary ones, and these again by tertiary ones, and so on. One example of this succession of organization centres has already been mentioned; in the chick there is first an organization centre in the endoderm which induces the primitive streak, which then becomes an organization centre itself and induces the neural plate. Many similar examples are known. In the newt, for instance, the neural plate closes up to form the neural groove, and the two sides of the groove eventually join together at the top and cut off the lower part as a tube sunk below the surface. From the front end of this tube, two hollow pro-cesses grow out, which are the beginnings of the two eyes. When these processes reach the surface and

touch the skin, they act as organization centres and
induce the skin to form a lens (Fig. 21). There are
other secondary organization centres which induce
various parts of the embryo to become legs or gills,
etc. We do not always know exactly which sorts of
tissue carry the organizing power, but we find that

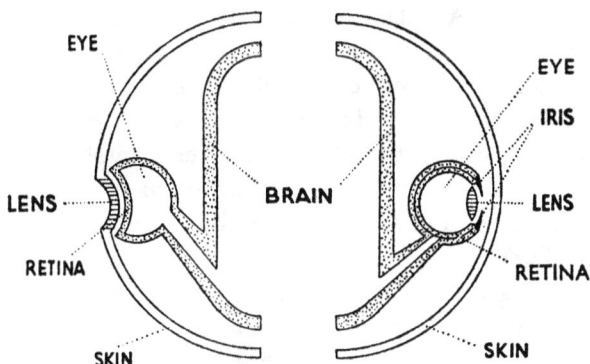

FIG. 21.—The development of the eye in vertebrates. A diagrammatic section through the head; the left side shows an
early stage and the right side an older one

if we graft non-presumptive leg material into the
region where the leg will appear, there is something
there which succeeds in turning this tissue into a leg.

Special Factories Making Special Parts

Once the various parts of the embryo have been
started off in definite directions by the organization
centres, they can continue developing on their own.
They are, in fact, capable of *self-differentiating*. This
does not mean that every single part of the leg, for
instance, has its fate absolutely fixed and can only

develop in one way. It means that if the lump of
presumptive leg tissue develops at all, it will develop
into a leg. But within the lump a good deal of
changing about of various parts can go on without
affecting the result. If a piece is removed the leg-
rudiment will nevertheless be able to form a complete leg. A system like this, where the various parts
can be changed round or taken away without
affecting the result of the subsequent development, is
called by the grand name of a *harmonious equipotential
system*. As a matter of fact, it is very doubtful if any
completely harmonious equi-potential systems exist.
There are nearly always some differences between the
different parts of the system so that changes of some
kind have no effect but others have. The leg-rudiment
may have a tendency to form a complete leg out of
whatever tissue is available, but the tissue is never
quite labile in every way and therefore, after some
injuries, cannot be moulded into an entire limb. But
the tendency to form a whole leg out of an injured
rudiment suggests that the tissue is, in some way
which we do not yet understand, in equilibrium
when it forms a complete whole. Dragomirov has
recently studied the question in the eyes of the newt
embryo. He cut out and isolated various parts of the
eye rudiment and watched how they returned to the
normal shape. He found that the isolated pieces could
develop in various ways, which represented different
short-cuts, as it were, back to the equilibrium position
of being a complete eye.

Once, when I was working in Germany, I met a

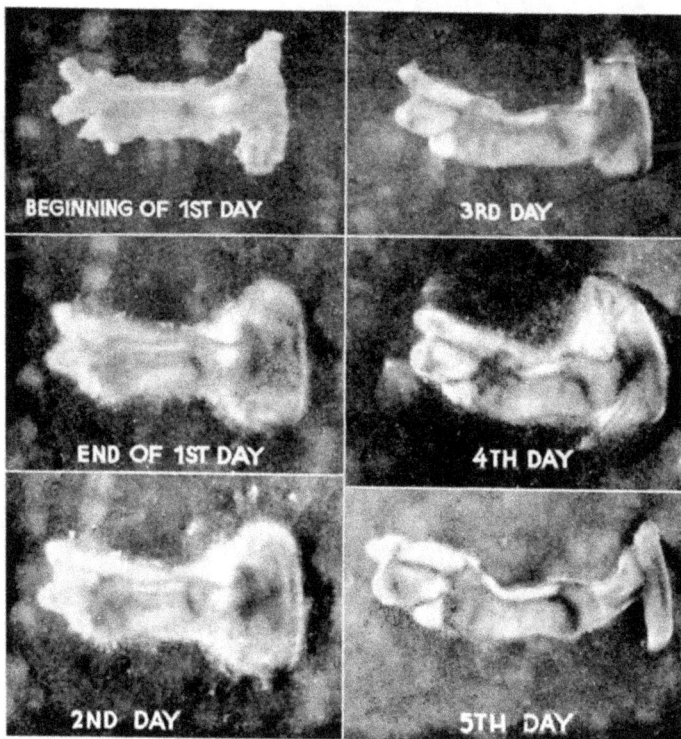

FIG. 22. —Self-differentiation of the leg bones of a chick in culture.
(From a cinema film by Canti and Fell.)

very enthusiastic Russian embryologist, who was full of the similarity between the organization centre and the Communist party. He said that the organizer arises as a small ordered part of the egg, and when a revolution happens (that is, gastrulation occurs) the organizer remains in control because of its own order and discipline, just like the Communist party in Russia, according to him. As soon as the main job is over the party starts a lot of separate factories going, each to perform some special function, and these factories are quite self-sufficient within their own limits and elect their own executives, just as the secondary organizers and the regions controlled by them can self-differentiate and act as harmonious equipotential systems, running their own show without much control from the primary organization centre. I do not know enough about Russia to say whether the Communist party really works like that, but the comparison does bring out some important features of the way the organizers work.

Once an organ rudiment has been determined and starts self-differentiating, it is amazing how fully and in what detail it develops. A very good example of this is seen if the rudiment of the leg-bones are removed from a chicken embryo and grown in a culture, like the cultures used for growing entire embryos. Miss Fell, collaborating with Dr. Canti, has made a cinematograph film of what happens, and the pictures shown in Fig. 22 are taken from this film. When the rudiment was isolated from the embryo, it was a little shapeless mass of mesoderm.

F

From this a well-formed leg had developed, and the cells had turned into cartilage cells and were beginning to turn into bone by the time the experiment ended. The ends of the bones at the joints of the hip and knee had nearly the right shape; as we shall see later (p. 111), the actual functioning of the joints and the movements which take place there have a certain influence on the shape of the articulating ends of the bones, but even without this help the rudiment can develop quite a good joint-surface. Further, not only did the bone-rudiment develop its right shape and right sort of cells, but the various chemical processes which go on in developing bone also occurred. There is a particular substance called phosphatase which appears at a certain stage or normal development and is concerned with the laying down of the calcium phosphate out of which the bone is made; and it was shown that the isolated rudiment had produced this substance at the right time.

Another very remarkable instance of the self-sufficiency of organ rudiments after they have become determined is found in the embryonic kidney or mesonephros of the chicken. This structure only exists in the embryo and degenerates and falls to bits before the chick is born. If the rudiment is cut out and isolated it goes on developing quite normally for the right length of time and then punctually starts to degenerate. A similar capacity for independent development has been found in the organ-rudiments of frogs and newts and most other animals which have been investigated.

Mosaic Eggs

The stage of development which has just been described, in which all the different parts of the embryo are determined, is called the mosaic stage, because then each separate part is independent and self-sufficient like the separate stones in a mosaic. In most embryos this stage is attained gradually, as the successive organization centres do their work. The primary organization centre cuts the embryo up into a mosaic made up of a few big blocks, the neural plate and skin, and so on, then the secondary organizers cut up those blocks into smaller ones, the eyes, ears, legs, and other organs. But there are some eggs, to which we have so far paid very little attention, in which the mosaic stage is reached very early in development, so that there is no time for us to perform experiments to find out whether there are any organization centres or not. In the most extreme cases the egg is a mosaic immediately after it is fertilized. In these eggs, of which the Ascidian, or sea-squirt, egg is one of the best known examples, we can sometimes show that there are definite substances each of which is essential for the formation of some particular organ. They are known as *organ-forming substances*. We do not know what they are, but in the eggs of some species of animals they have characteristic colours or textures so that they can be easily recognized. The essential fact about them is that if they are moved into an abnormal position in the egg, the particular organ into which they develop will be found to appear in this abnormal position: or if one such substance is removed from the egg altogether, the organ will simply fail to turn up at all.

If such a *mosaic egg* is cut in two, each half can only develop into half an embryo since it only has half the right organ-forming substances. This is quite different from the result of Spemann's experiment, which was mentioned in Chapter II, where it was described how he made one egg produce two whole embryos by constricting it in half with a hair. As a matter of fact, he was lucky in this particular experiment; in other cases he sometimes only got one embryo. Because, after all, the newt's egg has got one "organ-forming substance" or something very like one, namely, the organizer. Spemann found that only those halves which contained the organizer developed into embryos, and he got one or two embryos, according as the organizer was included whole in one half, or cut in two and divided between the two halves.

Different embryos can be arranged in a complete series between those which, like the newt, are late at arriving at the mosaic stage, and those which arrive early. In the former we can often prove the presence of an organization centre, but in the latter we cannot discover whether there is one or not. Of course in these mosaic eggs there may be an organizer which acts so early that we cannot find it, and induces the formation of the organ-forming substances. But it is impossible to explain all organ-forming substances by saying that they have been induced by organizers, because, if so, where does the organizer come from? It is rather the other way about. The ordinary newt organizer should be regarded as itself very like an organ-forming substance. Wherever it is, there the mesoderm develops out of it and the rest of the embryo is induced; if it is removed altogether, no embryo appears. In the frog's egg we can actually see the organizer, coloured grey and collected on one side of the egg where the blastopore will appear later, just like an organ-forming substance in an Ascidian's egg. The organizer must be prepared by chemical processes going on in the egg and then localized in one region by the external agents, which were mentioned earlier (p. 62). We know very little about the preparatory processes, but we have a very few hints about one kind of process which must be involved in the newt.

As we shall see in the next chapter, *one* of the important things about the organizer is that it contains an active chemical substance which gives the ectoderm the necessary stimulus to make it develop into neural plate. We do not yet know exactly what the substance is, though we are rapidly finding out. It is provisionally called the evocator, because it evokes the formation of the neural plate. Now it has been shown that an inactive form of the evocator is present in all adult tissues and, further, that the same inactive form is distributed throughout the whole newt's egg. In the organization centre this inactive form is converted, much earlier than elsewhere, into the active form which is found in the adult tissues. The preparation of the organizer therefore involves the activation of a substance. When we know more about the exact chemical formula of the evocator, we shall perhaps be able to find out how this activation happens.

The localization of the organizer, i.e. the determination of whereabouts in the egg the activation shall happen, must be done by factors outside the egg, partly by the way the egg is formed in the ovary and, in the frog, at any rate, partly by the place where the sperm enters the egg in fertilization. The localization of organ-forming substances in Ascidians and other mosaic eggs must also be done by external agents, and here again the same two factors are probably important.

Human eggs are not mosaic. They probably develop a primary organization centre which acts at the stage corresponding to that in which the second chick organizer is active, that is, at the primitive streak stage. The embryo reaches this stage about 8 days after conception. Before then, separate bits which may get split off by chance can develop into whole embryos if they are big enough. The most common sort of splitting is for the embryo to fall into two halves: probably after the egg has cleaved the first time the two daughter cells fail to stick together for some reason. We then get *identical twins*, which are not the same as *fraternal twins*, which arise when two eggs are fertilized simultaneously. Identical twins must, of course, have exactly the same hereditary factors, and if one can find identical twins which have been separated in early life and brought up in different families, one can measure the relative importance of heredity and upbringing by finding how much they do or do not resemble each other. So far only a fairly small number of pairs of separated identical twins have been carefully studied, so that the conclusions are not very certain, but, on the whole, the information which has been collected emphasizes the importance of heredity, even for shaping intellectual and emotional qualities.

Some animals have specialized in producing many embryos from one egg. One species of Armadillo, for instance, always has four identical quadruplets. Other animals, like some of the parasitic wasps which lay eggs in caterpillars of other insects, produce enormous numbers of young from one egg. In such cases the egg falls to pieces at an early stage in its development, and each piece then goes on to make a whole embryo. We may perhaps find out how it is done, and then we could control twin production at will, and produce a large number of babies from an egg with a particularly good set of hereditary factors.

Different Ways of Building Up Similar Structures

When an embryo reaches the mosaic stage the real elaboration of detail begins. The ways in which the different organs are built up are too manifold and too complicated to be summarized in such a short book as this. There is only room for a description of how the main outlines of the animal are laid down and an idea of the way in which this fundamental pattern is called forth. There are many excellent books, such as MacBride's *Invertebrate Embryology* and Graham Kerr's *Vertebrate Embryology*, where the details of the later processes of development are described. The reader who consults these books will be amazed at the mass of knowledge we have about the development of the embryos in different animal groups.

It has been said above that, as a rule, embryos of different kinds resemble one another the more the younger they are. This is not always true as regards the separate organs. It is very interesting to find that sometimes distantly related animals have evolved similar organs, but have evolved them in quite different ways which are recapitulated in their embryological development. The clearest examples of this are seen in organs which have to perform some quite definite function, which more or less dictates their structure. For instance, an efficient eye must have a sensitive layer of tissue (the retina) to perceive light, a lens to focus the light, and an apparatus like an iris-diaphragm on a camera to regulate the amount of light which enters. In the higher vertebrates, such as man, mammals, birds, and newts, the retina originates from an outgrowth of the neural plate in the brain region which we have already described. It induces the formation of a lens from the skin of the head. The iris is formed from the retina (Fig. 21). Now all these parts are quite well-developed in the eyes of octopuses, which belong to the group of animals known as Cephalopods and have hardly any connection with vertebrates. And we find, if we study the embryology of octopuses, that their eyes, although very similar in plan to vertebrate eyes, are developed in a totally different way (Fig. 23). The whole retina is made from a patch of *skin* which sinks inwards and

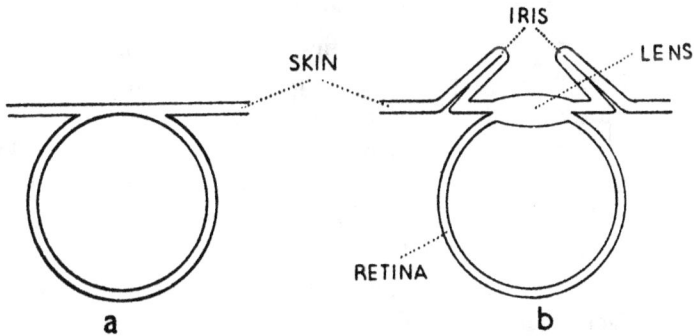

FIG. 23.—Development of the eye in the octopus. (*a*) Is a section of the hollow ball formed by a fold of the skin. (*b*) Shows the retina developed from this fold, and the iris from two more folds, one on each side of the eye

becomes cut off from the surface, so that it forms a hollow sunken sphere. Then the lens differentiates from the thin roof of this sphere, and two circular folds of skin grow up all round, the inner fold becoming the iris, while the other may join up across the top and become transparent and so change into a cornea, like the vertebrate cornea, which is a sort of protective eye-glass covering the pupil. The octopus-brain takes no part in the whole process, though it does, of course, eventually become connected to the eye by a nerve.

Larvae

In many animals the process of development is complicated by the fact that the animal passes through one or more larval stages before it becomes adult. The best known illustration of this is found in the life-history of insects. Everybody knows that a butterfly's egg develops first into a caterpillar and then into a pupa or chrysalis, and only turns into a butterfly after all this preparation. Some insects have a much simpler development; earwigs, for instance, hatch out of the egg as little earwigs, which look quite like the adult, although there are still a few changes to be made. But a caterpillar is extremely different from an adult butterfly: so different, in fact, that nearly all its organs have to be broken down and absorbed in the chrysalis stage, when they provide material which nourishes the new organs out of which the butterfly is built up. For instance, only the three pairs of legs nearest the head persist into the adult, and all the others disappear. The wings and most of the adult organs develop from little lumps of tissue called imaginal discs which are formed quite early in the caterpillar and wait till the pupa stage before they start growing. Fig. 24 shows the front part of a caterpillar of a swallow-tail butterfly, with part of the skin lifted up to show the imaginal discs from which the wings will grow.

Very many invertebrate animals and a few vertebrates go through larval stages before they become adult, and there may be a whole series of different stages, so that their life-history is very complicated. These larval stages often represent ancestors in the evolution of the animals. For instance, frogs were evolved from ancestors which lived in water, and

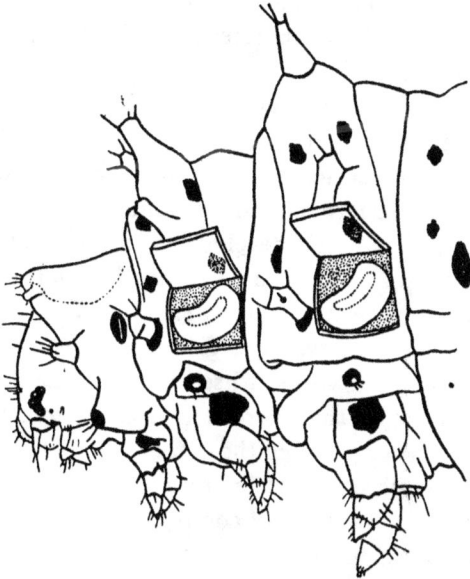

FIG. 24.—Part of a caterpillar with the skin dissected away to show the buds from which the butterfly's wings will arise

they still have a water-living larval stage, the tadpole. The Ascidians, which we mentioned before as examples of animals with mosaic eggs, are not so well known as frogs, but they also have a larval stage, which is rather like a tadpole. Here the metamorphosis by which the larvae turn into the adult is much more radical, although the adults remain aquatic. The tadpoles swim about freely, but the adults are fixed to the sea-bottom and are simple sacks, rather like sea-anemones in some ways. Nearly all trace of their original structure is lost, and we should hardly have realized that they are near relations of the vertebrates if we did not know what the larvae are like.

It is not very obvious why the life-histories of some animals should be complicated by the development of larvae forms when the animals are finally going to turn round and become something absolutely different. To develop into a butterfly by first becoming a caterpillar is rather like going to Birmingham by way of Beachy Head, as Chesterton has it. Probably it serves a useful purpose by making the embryo able to support itself before its development is complete, so that there is no need to store so much yolk in the egg to keep the embryo alive. But to show why something is useful does not explain how it arose. The evolution of complicated and specialized larvae may be a consequence of the competition between different young embryos for the food which is suitable to young as opposed to adult creatures. As soon as embryos start to feed themselves, they will be submitted to the struggle for existence and must prove themselves efficient or perish. Just as the same sort of competition led to the evolution of the adults, so it would force the embryos or larvae to become more specialized and complicated.

CHAPTER VI

THE DEVELOPMENT OF PATTERN

How the Organizer Works

THE discovery of organization centres is only a beginning of the analysis of why an embryo develops, but it is a very promising beginning. It provides some sort of an answer to the fundamental questions of embryology, why does a certain part of the embryo develop into this organ and another part into that? and how are the parts fitted together? We can already say why part of the gastrula becomes the neural plate. It does so because the organization centre stimulated it to develop in that way. But there are two reasons why that is not quite the whole of the story. For one thing, the embryonic tissues can only react to the stimulus of the organizer when they are ready to do so. The organizer does not seem to be able to induce anything in tissue younger than the early gastrula stage. We do not know why this is, but probably various chemical processes go on in the cleavage cells, so that by the time gastrulation begins they have become reactive and can be affected by the organizer. When they are reactive, they are said to be "competent," but the competence does not last long. After the end of gastrulation the ectoderm loses it power to form neural plates under the influence of the organizer. That is to say, it becomes non-competent again as regards this process; but as we have seen, it acquires other competencies and reacts to other organizers by forming other organs.

Although we know so little about competence, or what makes a tissue reactive, we have found out something about the stimulus which the organizer provides. It was first found, both in the newt and the chicken, that the organizer was still effective after it had been killed, so the stimulus cannot depend on any property peculiar to living cells. Quite recently it has been shown that the stimulus is really due to a chemical substance, but the substance has not yet been purified and analysed. When material containing this substance is implanted into an embryo, it stimulates the competent tissue to form an extra, induced, neural plate. But this neural plate seems to differ in a very interesting way from those which are induced by lining organizers. The usual induced neural plate is part of an embryo, it forms the nervous system of a head or a tail or part of the trunk (for instance, see the induced head in Fig. 18). The neural plates induced by the chemical substance on the other hand seem to belong to no particular part of the embryo; they are just neural plates with no definite shape to show which part of the embryo they are. If we think about it, this is what we should expect. Because it is clearly impossible for a single chemical substance acting on the competent ectoderm to produce all the different parts of the neural plate, unless, indeed, the various parts of the ectoderm differ from one another in some way, which does not seem to be the case. It seems, then, that the chemical substance simply gives the stimulus for the neural plate to be formed and that something else must be present in the living organizer to determine which part or parts of the neural plate it shall be. The chemical stimulus is called the *evocator* and the something else which determines which part of the neural plate shall be induced is called the *individuation field*.

The fact that an organization centre contains an individuation field is the second of the two reasons why it is not sufficient to say that Spemann's discovery merely tells us why some particular part of the embryo becomes neural plate. The full statement is that some of the ectoderm in the gastrula becomes neural plate because, while it is in a competent state with respect to this process, it is acted on by the evocator in the organization centre; and that it becomes a particular part of the normal neural plate fitted into the whole embryo because it is within the individuation field of the organization centre. The idea of the evocator expresses the way in which the organizer is active as a determiner, and the idea of the individuation field emphasises the way in which it is active as a centre of the organization of the embryo.

Particular Causes and the General Pattern

It is the individuation field which fits the different parts of the embryo together to make up one single complete animal. It determines what part of the

neural plate shall be induced by each part of the
organization centre. It therefore determines where
the secondary organizers like the eye-cup shall
arise, and then their individuation fields determine
where the tertiary organizers appear, and so on.
We cannot yet say how the individuation field gets
its effect, but we can describe what the effect is.
We find that the field is a region where there is a
tendency to build up one complete embryo, and
that if we add any extra material or take any away,
the material which is left will do its best, so to speak,
to turn into one embryo and not one and a bit. The
state of being one embryo seems to be a condition of
equilibrium to which the tissue tends to return.

Perhaps a mechanical analogy will make this
clearer without being, at the same time, too mis-
leading. Think of a big railway sorting yard, like
the one in Fig. 25. You are looking down an incline
called the Hump. The wagons are pushed over the
Hump and go running downhill and are sorted out
by the systems of points into the various sidings.
Now an embryo is in some ways analogous to a set
of trucks sliding down the Hump. The first point,
which you see just in front of the nearest two trucks
in the picture, is the primary organization centre
and shunts off one set of trucks to the left, to become
skin, and another set to the right to become neural
plate. The next set of points are the secondary
organizers, which we discussed in the last chapter,
and they again sort out the neural plate trucks into
brain trucks and spinal column trucks, and the

FIG. 25.—Whitemoor Marshalling Yard, L.N.E.R.

skin trucks into lens and epidermis trucks. And so on through all the sets of points, representing all the organizers, till the final trains are made up in the sidings, or, as we may say, the final organs are developed from the embryo.

In this model a competent tissue is analogous to a truck just approaching a point. According as the tissue is acted on or not acted on by an organizer, it develops into this or that, and according as the truck finds the points this way or the other, so it runs down this siding or that. But in this comparison each point would act as an evocator and not as an organizer, unless there is something to compare with the individuators of the organizers. Individuation by the primary organization centre means determining what detailed structure the induced organ will have, it means settling whether it will be brain or spinal column, etc. In our model, therefore, it means sorting a whole set of trucks into their final sidings so that each siding gets the right number. The first point, or primary organization centre, can only do that if it is connected in some way to all the other points by a system of levers so that they all work together in a co-ordinated way. The points alone do not provide a full analogy with living organization centres, one must bring in the set of individuating levers.

The important thing to remember in this analogy is that evocation, or the action of one point, can happen to one single truck, that is to say, to one small piece of tissue. But individuation, the harmonious sorting out of a train of trucks, can only be seen if there are a lot of trucks, that is to say, it can only happen to a mass composed of several different regions of tissue. The essence of individuation is to combine several different parts harmoniously together so as to form a complete organ. If a piece of a neural tube is induced, whether we recognize it as a brain or as part of a spinal column, depends on how the top and sides and bottom are shaped and fitted together. The actual organizers we know of always fit the various parts together as harmoniously as possible. We do not find the walls of a brain fitted together with the top and bottom of a spinal column. This tendency to make a complete unit is the most important characteristic properties of the individuation field, as opposed to the evocator.

There is one other possibility which is also well expressed in this analogy. In the real sorting yard, the building up of the right trains is not done by a set of compensating levers, such as we have imagined to account for individuation, but instead it is done by a man sitting in the control-box and using his brains. Many authors, the most famous of whom is Driesch, have suggested that we cannot imagine any material compensating system in the embryo which would be efficient, but must postulate a non-material agency which works just like the man in the control tower. This agency is called an *entelechy* and is supposed to be responsible for the harmonious development of the different parts of the embryo. It is non-material and cannot be directly discovered by experiments. It is not supposed to have any connection with the religious or moral soul, but it is a sort of developmental soul. A hypothesis like this is in the first place a confession that we do not understand how the harmony of development is brought about, but more than that, it denies that we shall ever find out. It is frankly defeatist, and on those grounds alone it is usually discounted by scientists at the present day, who feel that, although they do not understand all about development yet, they still have no grounds for saying that they never will understand it in the future. Moreover, the organizers have an easier job than the pointsman in a shunting yard, because in any one kind of embryo the final state is fixed. It is as though there was always the same set of trains to be made up in the final sidings. The organizers have only to deal with variations in the raw material, while the pointsman has as well to build up different sets of trains on different days.

The living newt organization centre therefore does two things, it evocates a neural plate from the gastrula ectoderm, and then it individuates itself and the neural plate and its surroundings generally into a complete embryo, or as big a part of one as possible. And we can separate these two processes by using dead organizers, which evocate but do not individuate. Some of the other organization centres we know of seem only to individuate. This is so with the sea-urchin centre discovered by Hörstadius; wherever the small vegetative cells are placed they try to build up a whole embryo. Perhaps Seidel's activation centre in the dragon-fly can be regarded as an organization centre which evocates but does not individuate, whereas the differentiation centre does individuate.

Regeneration

In many animals individuation fields persist all through life. Their effect is seen when such an animal is badly wounded, loses a leg, for example. The stump of the leg completes itself and builds up a complete leg again by regenerating the part which has been lost. In this regeneration the processes which go on are often very similar to those connected with organization centres in earlier embryonic life. For instance, if an adult newt loses a leg, a little cap of cells called the *regeneration bud* grows on the end of the stump. The regeneration bud is simply a little bit of competent tissue, and the individuation field controls it so that the lost part of the leg develops out of it. It is not only competent to form legs but can turn into other things if necessary. If it is taken off and grafted on to the stump from which part of the tail has been cut, the tail individuation field now uses the bit of competent tissue to complete *itself*, and the regeneration bud, instead of becoming part of the leg, becomes part of the tail.

In many of the lower organisms, all the tissues seem to be permanently in a competent state, but they are normally held together as a complete individual in an equilibrium which is determined with respect to some particular part, which is usually the head. If the animal grows so big that part of it gets out of range of the head's influence, this part may develop on its own into a whole new animal, and we thus get the well-known phenomenon of reproduction by budding, which happens in corals, for instance. Or if the head is cut out of one animal and grafted into another some distance away from the host's head, it can overcome the influence of the host's head and start to build the competent tissues in its surroundings into a complete new animal. These facts can be very easily described in terms of the hypothesis that there is a gradient of something along the main axis of the animals, with the head to the top end of the gradient and the tail at the bottom. The hypothesis suggests that when any part of the gradient is isolated from the rest it tends to change itself back into the whole arrangement, and the head or top end is particularly well able to do this because it can alter any lower part which may be in its neighbourhood so as to bring that part into conformity with itself. Gradients of this kind are called *Axial Gradients*, and attention was first drawn to their importance by Child. They may perhaps always occur in individuation fields in some form or other. But until we know rather more about what they are gradients of, they must be regarded more as a descriptive convenience than as a real explanation.

CHAPTER VII

THE FINAL ADJUSTMENTS

THE development of an animal is not finished as soon as all its organs have appeared. The details of its structure must still be fitted together and adjusted to one another. Organs like the heart, which are required early in embryonic life, get a start on other less necessary parts and are disproportionately large in the young embryo. The control system of nerves has to be connected up. Particular parts may have to be enlarged to meet special demands made on them. In this chapter some of the ways in which the final "tuning" is carried out will be briefly described.

How Embryos are Fed

The embryo has to be provided with food until it is old enough to catch and eat its own nourishment. The study of the food-supplies of embryos, and the changes by which the food is digested and then built up into the tissues of the growing animal, is a very complicated one. Until recently the whole subject was a chaos of unrelated facts, but a few years ago Needham brought together all the scattered pieces of information in a work entitled *Chemical Embryology* and stated the general principles which emerged. In this book we are mainly concerned with the development of animal shape, and there will only be space for a short outline of the most important results of the chemical study of embryos. The most interesting point of view from which to consider the question is that of evolutionary adaptation; the different environments in which animals live make different demands which must be satisfied by animals which live in them.

The sea, where life first evolved, does not provide many of the substances necessary for a developing egg. These substances are (1) oxygen, (2) water, (3) inorganic salts, (4) carbohydrates, like sugar, which can be burnt to provide energy, (5) fats, also mainly for fuel, (6) proteins, which are the real "flesh-forming" substances. Of these the sea provides the first three. The others have to be supplied to the embryo in the yolk. But the sea abounds in very small living creatures which provide suitable food for quite tiny larvae, so that the amount of yolk need only suffice for the earliest stages of development.

The freshwater environment does not contain enough salts to be absorbed by the embryo, and it therefore can only be inhabited by creatures which provide sufficient salts in the egg. But the most difficult evolutionary step, from the embryological as from every other point of view, must have been the conquest of dry land. Here there is no useful substance, except oxygen, which can be absorbed from outside, and the embryo has to be provided not only with its organic materials and salts but also with its water. Amphibia, which are the most primitive land animals, have, as a matter of fact, failed in the last respect, and must return to water to breed since their embryos would otherwise dry up. The first true land animals were the reptiles. They and their descendants, the birds, are quite independent of water for breeding, as they have evolved the device of enclosing the egg in an impermeable membrane which does not allow the moisture to escape, although it allows gases such as oxygen to be absorbed. Some of the intermediate stages in this evolutionary advance have still survived; for instance, one sort of turtle lays eggs with a rather soft and permeable covering, which can only develop if they are deposited in wet sand; the turtle scrapes a hole in the sand, lays the eggs, and then urinates over them to moisten them.

The closed-box, or "cleidoic" egg, although it solves the drought problem fairly satisfactorily, is only workable if several other requirements are fulfilled. The most important of these concerns the waste products, of which the embryo, like all other animals, has to rid itself. One waste product is carbon-dioxide, which is gaseous and can be allowed to diffuse out through the shell. The other main type of waste is the nitrogenous matter derived from the breakdown of proteins. Water-living embryos turn this into ammonia which diffuses out into the surrounding water. But it would not diffuse away so readily from a closed-box egg and would accumulate to such an extent that it would be dangerous. The closed-box animals have therefore had to evolve another method of getting rid of their nitrogenous waste products. What they actually do is to convert the nitrogen into fairly insoluble substances, such as urea, or, in some highly evolved animals, uric acid, which is kept in a special organ until the embryo escapes from the shell and can excrete it. In the development of birds, for instance, there is a very striking example of chemical recapitulation in this respect; in the early stages the nitrogen is excreted as ammonia, as it is in fish, then it is converted into urea and finally, at the end of development, most of the waste nitrogen is being formed into uric acid. Birds and reptiles have also become modified to produce less nitrogenous waste; they do this by using more fat and carbohydrate and less protein, as fuel, and they get the added advantage from this that the fat and carbohydrate give rise to some water when burned and thus help with the provision of the necessary moisture.

So far nothing has been said about the feeding of the mammalian embryo. This is because of the quite special and peculiar conditions which are involved in uterine life. In a sense, the mammals have found a way of returning their embryos to the sea, the "sea" in this case being the maternal blood, which can supply oxygen, water, and salts, and can remove waste products, and which has the additional advantage of supplying carbohydrates, fats, and proteins as well.

The essential feature of the evolutionary step taken by the mammals is to place the embryo inside the body of the mother and to bring the embryonic and maternal blood circulations into close contact. In some other animals, for instance, some kinds of fish, the embryo develops for some time within its mother's body, but it is only in the mammals that the connection of the two blood streams is at all intimate. Even in mammals maternal blood never circulates actually through the embryonic blood-vessels. If it did, not only might the mother's female sex hormones disturb the development of the sex of her sons, but the embryo would actually be killed by the foreign proteins, since each animal has a chemical individuality which has to be respected. There is therefore always a filter between the maternal blood and the embryo. This filter is the placenta, which is developed from the trophoblast, or non-embryonic part of the embryo which was described above. The trophoblast lies against the wall of the uterus, and substances which pass from the mother into the embryo have to pass through the lining of the uterine blood-vessel, through the connective tissue surrounding it, through the wall of the uterus, then into the embryonic circulation through the wall of the placenta, through connective tissue, and finally the lining of the embryonic blood-vessel. There are thus six layers to be passed through (Fig. 26, a). But these conditions only hold in the most primitive types of placenta; they are found, for instance, in the pig and the cow. In some other mammals the passage of substances is made easier by the disappearance of some or all of the maternal layers. In the stag the maternal wall of the uterus disappears over large areas, in the cat the uterine connective tissue goes, and in mice, rabbits, and men the walls

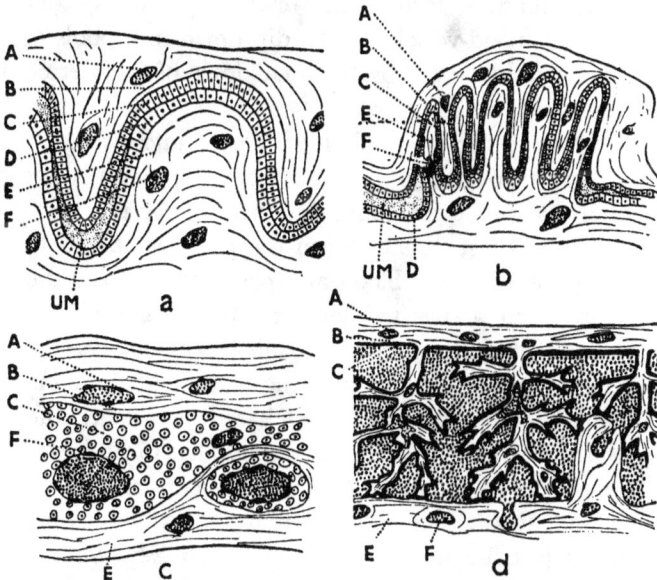

Fig. 26.—The different types of placenta. The part derived from the embryo is above, and that derived from the mother below in each figure. (*A*) The lining of the embryonic blood-vessel. (*B*) Embryonic connective tissue. (*C*) Embryonic membrane. (*D*) Maternal membrane. (*E*) Maternal connective tissue. (*F*) Lining of maternal blood-vessel. In (*a*), which is the least developed type, all six layers are present. In (*b*), layer (*C*) lies against (*E*) in the folds. In (*c*), (*D*) has disappeared entirely and (*C*) lies against (*F*). In (*d*) (man), (*C*) lies against the maternal blood itself. (*UM*) is the uterine milk in (*a*) and (*b*)

of the uterine blood-vessels break down and the maternal blood lies right against the wall of the placenta, and there are only three layers left separating it from the embryonic blood. In the same series of animals the wall of the placenta becomes more and more folded so as to increase the area over which the absorption of food can take place. In man the placental wall forms thin branching fingers, containing blood-vessels, which dip down into pools of maternal blood left by the breakdown of the uterine blood-vessels.

Most of the embryo's food passes into its blood stream through the placenta, but the process is not really a simple filtration. The wall of the placenta, or trophoblast, breaks down the proteins offered to it by the maternal blood and passes them on to the embryo in a digested form ready to be built up again into the particular substances which the embryo requires for its growing organs. At the same time the placenta passes back to the mother the waste products of which the embryo has to rid itself; in the more elaborate types of placenta this backward passage is so efficient that the embryo never develops a functional embryonic kidney, which other embryos have evolved to deal with their excretion before the final adult kidney is ready.

Not quite all the embryo's food comes from the mother's blood. Some is provided by the so-called uterine milk, which is secreted by glands in the uterus, chiefly in animals with simple placentas. In the more complicated types, the embryo derives a certain amount of nourishment from the maternal tissues—the wall of the uterus, the connective tissue and the lining of the blood-vessels—which it digests and absorbs. This process goes on particularly rapidly in early stages of development. In man, young embryos are very difficult to find, and as yet no stage is known which has not burrowed into the side of the uterus, digesting the maternal tissue so as to prepare a space in which to develop embedded in its mother's flesh. The embryo must start on this cannibal existence almost immediately after it is fertilized. In fact, the connection between the embryo and its mother is so highly developed and specialized in Man that no further evolution along these lines seems conceivable.

It is often the failure of such a specialized organ to be able to evolve to meet new conditions which leads to the extinction of a species, and some authors have suggested that Man may eventually die out because of the over-specialization of his placenta.

Growth

Many animals never stop developing their whole life long. It is only for convenience that we say that a man becomes adult at the age of twenty-one. And similarly with other animals, there is often no point at which we can say definitely that the period of change is over. This is particularly the case with animals which go on growing throughout their whole existence. With others, such as butterflies, once the pupa has metamorphosed into a butterfly, no more change is possible, and there can be no objection to calling the animal adult. All animals which grow, change not only in absolute size, but also in proportions and therefore in shape. This alteration is most rapid, of course, in the early years of life and gets slower and slower as the animal gets older. Fig. 27 shows how much the proportions alter as a baby grows up to be a man. There is often an evolutionary significance in these changes of proportion and we can show, in the evolutionary tree of

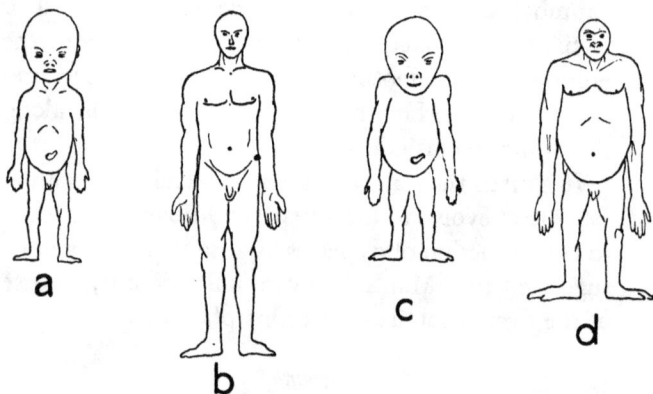

Fig. 27.—The changes of proportion during growth. (*a*) and (*c*) are human and gorilla foetuses, (*b*) and (*d*) an adult man and gorilla. All the figures are adjusted to have the same sitting height. Notice that in man the upper part of the body retains more nearly the foetal proportions, whereas in the gorilla the legs remain more unchanged while the head, and particularly the brain-case, becomes very small. (After Schultz.)

the horses, for instance, which we know from fossils, how one species has changed into another by an alteration of the rates of growth in different directions. Professor Huxley has recently written a fascinating book on this subject called *Problems of Relative Growth*.

The rates of growth of the different parts of a young animal or embryo at any given stage are not quite rigidly fixed, but can be altered for two reasons: either so as to fit in with its surroundings, or in response to the amount of work the part has to do. An alteration due to the first reason is seen if a small lens is grafted into a large eye: the small lens grows faster than it normally would and the large eye more slowly, so that together they make up a harmonious but middle-sized organ. The influence of the second factor, the amount of work required, is shown by a comparison of tadpoles which have been reared on a rich diet of animal matter with those reared on a poorer vegetable diet. The gut of the vegetarian tadpole has to work harder to absorb enough nourishment from its poorer food and grows much longer than that of the flesh-eating tadpoles (Fig. 28).

The Influence of Function

The work which an organ has to do and the movements it makes may influence not only its size but also its structure. There is a very good example of this in the differentiation of the leg-bones of the chick. It was found that the isolated thigh-bone can produce quite a well-shaped end to fit into the knee-joint even when it is lying in a culture with no muscles attached to it and makes no movements at all. If a thigh- and shin-bone are grown together in this way, both sides of the joint begin to form, but soon the two bones fuse together and the joint becomes obliterated. If, on the other hand, the two bones are exercised (by waggling them by hand), the joint goes on developing properly. Thus movement seems to be very important for the development of a really satisfactory joint.

The movements which developing embryos make are not the only functional factors which have im-

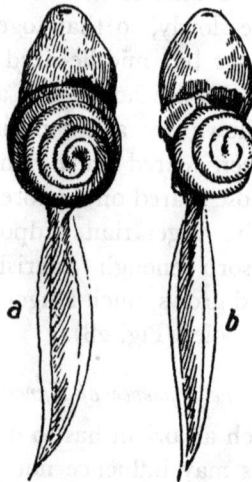

FIG. 28.—Differences in the guts of tadpoles fed (*a*) mainly on vegetables, (*b*) exclusively on meat. (From Dürken.)

portant effects on development. Sometimes a developing structure is put under tension because it is pulled out by its rapidly growing surroundings. Such tensions are probably very important, because if anything is stretched, any minute particles within it tend to become orientated parallel to the direction of stretching. Now it seems that the spaces in the embryo are sometimes filled with a jelly-like substance which is made up of little elongated particles which get orientated in various ways under the

(a)

(b)

FIG. 29.—Effects of the pituitary gland. (a) Giant
with over-developed pituitary. (From Engelbach.)
(b) Dwarf with under-developed pituitary. (From
Behrens and Barr.)

tensions set up in the developing embryo. Many processes of development involve the movement of cells from one place to another inside the embryo, and we know that cells will often move along lines of orientated particles. In this way the tensions may lay down the paths followed by the streams of cells and thus determine where they arrive.

The Nervous System and Development

The nervous system, which is so important for the integration of the adult body, might be expected to play an important part in enabling the embryo to develop harmoniously. Actually it has been found to be less effective in this particular respect than one might suppose, but many other interesting points arise about it. Before the nervous system can integrate the parts of the body, the nerves must grow out from the neural tube and reach the organs which they finally innervate. While the nerves are growing the paths along which they move are fixed by the positions of the organs to which they will become attached. If a limb-rudiment is cut out of its normal situation and grafted a short distance away, the nerves bend out of their usual course to follow it. The developing leg seems to attract them. This attraction can be exerted by most sorts of developing tissue, and if a leg is cut off and a nose grafted nearby, the leg nerve fibres cheerfully grow off towards the nose. But the actual way in which the nerve joins on to the leg muscles or the nose is another matter, and is more rigidly fixed.

H

Once the nerve fibres have got to their destination and become attached, there is a mutual interaction between the nerve and the organ. If the organ is made smaller than normal in any way or is removed altogether, the number of sensory nerve fibres coming back from it carrying sensations up the spinal column to the brain is reduced, although the number of motor nerve fibres going down the spinal column to carry orders to the organ is not affected. On the other hand, the nerve sometimes affects the organ; thus it is usually impossible for an organ to regenerate if the nerve is cut. In view of this dependence of regeneration on the presence of nerves, it is surprising to find that nerves are quite unnecessary for ordinary embryonic development, and that it is possible to prevent any nerves at all from growing into a leg-rudiment without this having any very great effect in the development of the leg, except that the muscles, although well formed, do not function and therefore do not grow as large as usual.

Hormones

Much of the integration of the developing animal seems in fact to be done not by the nervous system but by substances circulating in the blood-stream. It is usual to call these substances *hormones*, but it is difficult to give an exact definition of this term. They are usually thought of as substances which are present in the blood in very small quantities and which have a specific effect on various organs, but, as we shall see, some of these effects can be brought about by other non-specific substances. The hormones are usually manufactured in special organs set aside in the body for that purpose, which are called the *ductless glands*, because they have no opening to the exterior of the body but pour out the substances they make directly into the bloodstream. The ductless glands have all sorts of different origins in development. For instance, the pineal gland in man is formed from a structure which in our remote ancestors was a third eye situated on top of the head, and in its embryonic development traces of this ancestry can still be found. One of the most interesting of the glands is the pituitary, which is built up partly from the ectoderm of the mouth region and partly from the floor of the brain. It lies well hidden in the skull, underneath the brain, and seems to manufacture several different substances, all of which are very active and important. One of these substances has a great effect on growth. If it is produced in too small quantities in a child, that child becomes a dwarf, whereas if it is produced in too large amounts, the child becomes a giant (Fig. 29). Sometimes it is only in adult life that the gland starts to produce too much of the growth-controlling hormone, and then the main effects are seen in the face, hands, and feet, all of which become very large and coarse, giving the condition known as *acromegaly*. Other effects of abnormal amounts of hormones are shown in Fig. 30.

Hormones play an important part in animals

which pass through a larval stage. In the frog it has been shown that it is a hormone which makes the tadpole metamorphose into an adult frog. The hormone in question is produced by the thyroid gland, which lies in the neck in man, and gives rise to a form of goitre if it becomes too big. It is interesting to find that human thyroid hormone can be effective in causing metamorphosis in frogs, so the hormones are not special for each species of animal. It is probable that the moulting and metamorphosis of insects is also controlled by hormones, but so far no vertebrate hormone has been found which will produce the effects.

The chemical constitutions of some of the hormones have already been worked out, and the extraordinary fact has been discovered that it is sometimes possible to substitute for a natural hormone various other substances which are chemically rather unlike it. Perhaps the most striking example is in one of the numerous hormones concerned with regulating the development of the sexual organs. A whole series of substances are known which can be substituted for this hormone, which is called *oestrin*. All these substances have certain points in common, although they are quite dissimilar in other details. Moreover some of them can do quite other things; some can act like the vitamin D, in preventing animals getting rickets; some of them can cause cancer to develop if they are painted on to the skin constantly for months; and some of them can actually act as evocator-substances and cause the induction of

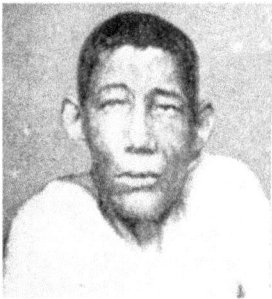

FIG. 30.—Comparison of the faces of various sorts of dogs with those of men suffering from hormone diseases. (From Stockard.)

neural tubes in newt embryos. It seems as if they were a group of skeleton keys each of which can unlock several different doors.

Sex

The development of sex is interesting not only in a general way, because sex is such an important element in our make-up, but also from the strictly embryological point of view it raises quite special and remarkable problems. When an animal develops, not only must the sex-cells, eggs and sperm, be elaborated, but the animal assumes a particular sexual character which affects its whole constitution, making it either male or female. These two phases of sexuality are to some extent distinct, at least in the way they arise.

The most prominent fact about the sex-cells is that they are capable of starting off a new development into another organism. One fertilized egg produces, among other things, another egg which can be fertilized and so on *ad infinitum*. This behaviour was emphasized in the Theory of the Germ-Track, which states that there is a continuous succession of germ-cells, the later ones derived from the earlier ones by division and reunion in fertilization, so that they are really all parts of the same living substance. This line of germ-cells represents, then, an *immortal* piece of living matter, which may increase in size but need never die. The thing which does die is the individual animal body which the germ-cells make to provide themselves with a temporary house.

One can only make a rigid distinction of this kind between the temporary body and the immortal germ-cells if one can show that the germ-cells are never really a true part of the body but are always fundamentally separate from it. Now in many cases it is probably true that the germ-cells never do belong to the body in any sense except that of just being inside it. We find, for instance, that at the very beginning of development there may be a particular part of the egg which looks unlike the rest and is set apart to become the germ-cells; in such cases there is, in fact, an organ-forming substance for germ-cells, and the rigid distinction between body and germ-cells may be plausible. But often no such substance can be found, and the germ-cells are probably formed in the same way as the other differentiated cells in the body, that is, in response to some organizer. In this sort of embryo the germ-cells really act like a part of the body and not as though they were inhabiting it and did not belong to it.

Then again, it is not only the germ-cells which are potentially immortal. All young embryonic cells are, and probably even the highly specialized cells in the different organs of an adult animal could remain alive almost indefinitely even although they might not be able to grow and divide. It is the animal as a whole which dies, not the individual cells of which it is made. The immortality of the cells can be realized by removing them from the body and growing them in a nutritive medium in culture, like the embryos mentioned in Chapter III. Anyone who wishes to can now make certain of the immortality of part of himself as an actual material living thing by endowing a staff of experts to keep the cultures going. Each culture, of course, only contains a very small piece of tissue, a few hundredths of an inch across, so that it is only rather a piecemeal immortality which science offers. But at the same time each of these little cultures is growing, doubling its volume in 2 days, and the tip of your little finger would grow to the size of the world in less than 6 months.

We must now turn to the second question; what decides whether the germ-cells, and the whole animal, shall be male or female? Actually two groups of factors work together in this sex-determination, and sometimes one is the more important, sometimes the other. One set of factors are given by the hereditary make-up of the embryo and the other set are particular conditions which arise during development. The hereditary basis is probably always present even where its influence is overcome by the factors which arise later. The important thing to notice about the hereditary determination of sex is that all animals inherit tendencies towards both maleness and femaleness, sometimes with one preponderating, sometimes with the other. The hereditary factors or genes for maleness and femaleness are usually carried on different chromosomes. One scheme is to have all the genes for one kind of sexuality, say femaleness, carried on one pair of chromosomes, which are then called the X-chromosomes, while the factors for maleness are carried on the other pairs, which we can call the A pairs to distinguish them. Then the sex which actually develops in an animal depends on the proportion between the Xs and the As, and we find that a special mechanism is evolved to regulate this proportion. Thus if the Xs are the female-bearing chromosomes, a female animal is found to have two Xs, making a normal chromosome pair, while the males have only one X, the other X being either completely absent or represented by a special "empty" Y chromosome, which does not carry any sex factors. The females therefore have a proportion of two Xs to the set of As, while the males have one X to the set of As. This plan makes certain that in the next generation the males and females will be about equal in number. All the eggs must contain one X and one of each of the A pairs, but the sperm will half of them contain an X and half no X (or a Y if the animal belongs to a kind which has them). In fertilization there will be equal numbers of animals which get an X in both the egg and the sperm and of animals which get only an X from the egg and none from the sperm. The first kind will be females and the second kind males. Sometimes, however, something goes wrong and the chromosomes do not divide properly when the eggs and sperm are formed, and in this way animals may get too many Xs in proportion to their As, when they develop into so-called "super-females," or again there may be too few Xs when we get "supermales"; and finally we may get proportions intermediate between one X and two Xs to the set of As and then intermediate forms or "intersexes" develop. Again, there are strong and weak varieties of the male and female factors, and unbalanced intermediate animals can result from some combinations of them. Most of these abnormalities have been found in insects, though they may perhaps occur in other groups where they have not yet been discovered. They are nearly all infertile and generally feeble.

The sex genes probably work by producing special sex substances or hormones, and if a young embryo comes in contact with these substances before its own sex genes have had time to act, the substances may get in first and overcome the genes. Or again, the two influences may act antagonistically and produce intermediate or intersexual animals. Even in adult life the sex hormones can produce very important effects. By various experimental methods it has been possible to bring about the complete reversal of sex in some animals, so that, for instance, a hen which can lay eggs may be changed into a cock which is functional as a male. The interaction of the hereditary sex factors and those which arise during development obviously can lead to very complicated results, and there is no space here to discuss the matter at all fully. Those who wish to go more deeply into the matter should read Professor Crew's book, *The Genetics of Sexuality in Animals*.

The Part Played by the Genes in Development

Finally, now that we have got an idea of how development comes about, we must go back to the question which was raised in Chapter II. How do the hereditary factors or genes affect development? As a matter of fact we have very little definite experimental evidence on this matter, because as bad luck will have it, the most suitable animals for embryological investigations, such as the amphibians, are just those animals about whose heredity we know least, because they breed so slowly and are so difficult to rear to sexual maturity. So all we can do is to make a few guesses.

In the first place, we know rather more about the effect of genes on the last, functional period of development, which has been the subject of this chapter, than about their effects on the earlier periods. In this last period, all sorts of details are being worked out, and sometimes it is not difficult to find out what factors are involved. For instance, if pigment is being made, in an animal's eyes or hair or in a flower's petals, we know more or less what chemical processes are going on. We find that genes affect such processes in at least two obvious ways, either by altering the quantity of some substance, or by altering the rate at which the reactions go on. Probably both these effects are actually due to the same sort of cause; different genes cause different amounts of special substances to be produced, and some of these substances may have an effect on the rate at which the processes proceed.

We find evidence which suggests, though it does not prove, that the genes act in a similar way in the more important earlier periods of development. For instance, we find that when an organ is induced by an organizer, the details of its shape depend on the tissue itself and not on the organizer. If we graft frog's skin in the mouth region of a newt, it becomes induced to form a mouth, and then it forms a frog's mouth, which is totally unlike a newt's. This means that the organizer determines that it shall be a mouth but that the competence determines the particular kind of mouth. The competence of the tissue, then, is always limited by the characters of the particular species it belongs to. Now these specific characters are, we may suppose on analogy with other animals, themselves fixed by the genes. If this is so, but we cannot yet prove it, then the processes which bring the tissue into a competent state must be controlled by the genes.

In all these reactions the genes are not acting alone, but are co-operating with the living matter outside the nucleus, or cytoplasm. We can only discover genes by breeding experiments in which we cross different varieties of animals, so that we only know how genes produce differences in development. Probably the cytoplasm provides much of the fundamental mechanism by which development is brought about, and the genes act as directing and controlling agents. The cytoplasm, on this view, would be, as we might say, the drilling-machines and lathes with which the animal is made, and the genes would be the particular tools, jigs, and drills which are fitted on to the machines for the actual job on hand. This idea may seem to suggest that the real fundamental thing is the cytoplasm and that the effects of the genes are purely superficial. To some extent this may be true for any one embryo. But one must not forget that the cytoplasm itself has been evolved through a long series of ancestors and has probably been affected by the genes which they contained. To pursue our analogy, the drilling-machines and lathes were shaped by drills and bits and jigs when they were built. Of course, to insist on pursuing the argument *ad infinitum* leads to a ridiculous question, like asking whether the hen or the egg came first, because finally the gene and the cytoplasm are dependent on each other and neither could exist alone.

Can genes affect organization centres? The answer seems to be "yes." True, we know very few cases where genes seem to have a direct effect on an evocator. In amphibians the evocator is present in the egg long before it is fertilized and cannot be manufactured by the genes contained in the fertilized egg, though perhaps it may be influenced by the genes of the egg's mother. In the insect egg Seidel showed that the activation centre, which is a sort of evocator, is formed by a reaction between cytoplasm and nuclei, and maybe it is the genes in the nuclei which take part in the reaction; but we do not know this for certain. In one species of fly, also, we know of a gene which makes a wing grow in the place of one of the other appendages, or a foot instead of the antenna, and it certainly seems as if this gene must have altered the evocators.

There is more evidence that genes can affect individuation fields; our original example of the short-finger gene in man is an example, since clearly this abnormality involves an alteration of the individuation field of the hand rudiment. In fact, all genes which affect patterns, even if they are only patterns of coloured hairs on the skin of a dog, may be said to affect an individuation field. The differences between the races of dogs, whose faces are shown in Fig. 30, are probably, in some cases certainly, caused by genes, and we may say that these genes have altered, though rather slightly, the individuation fields or patterns of their faces. The analogy with the human faces suggests that this alteration has been done by changing the amount of various hormones in the blood-stream, which have then caused alterations in the growth-rates of different parts of the face. It may seem that this has solved the problem, and has shown that an alteration of pattern is only a special case of an alteration of the quantity of a substance which is produced. This may be correct in some cases though in others the alterations produced by genes are more radical and can hardly be explained so simply. Such an explanation cannot, however, dispose of the problem of how patterns arise. Because we have still got to say why different parts of the face are differently affected by an alteration in the hormones. There must be a pattern of sensitivity, otherwise the face would just get bigger or smaller as a whole. In many cases it seems certain that the genes affect the fundamental patterns of sensitivity and do not merely cause one and the same pattern to be expressed in different ways.

SUMMING UP

This question of the pattern of the individuation field is the root of the whole matter. We have seen how the mere description of the patterns which are found led to the Recapitulation Hypothesis and the discovery of the fundamental pattern of the three Germ Layers. And we have seen in outline how the elements of the pattern arise. Processes go on in the egg which produce organ-forming substances, or evocators and competent tissues which react to give determined organ-rudiments. But how are the organ-forming substances or organ-rudiments arranged to form the pattern of a complete harmonious embryo? Somehow they must be localized. It is not difficult to imagine one or two substances being localized; it might be done by the way the blood-vessels bring substances up to the egg while it is developing, or it might be done (and, in fact, sometimes is done) by the sperm when it enters the egg at a definite place. In such a way we can account, say, for the localization of the evocator in the newt's egg. But how shall we account for the localization inside the organization centre of the factors which make part of the neural plate become brain and another part spinal column? We do not know. Probably we shall have to invent quite new methods of attacking the problem before we shall solve it. It may be that the structure and orientation of the minute particles of the cytoplasm are important. But it remains a problem for the future. The discovery of organization centres, to which so much of this book has been devoted, only clears the air and makes it possible to formulate more definitely the problem of what causes the patterns in which animals develop and on which their adult functioning depends.

INDEX

For Product Safety Concerns and Information please contact our EU
representative GPSR@taylorandfrancis.com
Taylor & Francis Verlag GmbH, Kaufingerstraße 24, 80331 München, Germany